MINITAB® MANUAL

The Triola Statistics Series:

Elementary Statistics, Eleventh Edition

Elementary Statistics Using Excel,
Fourth Edition

Essentials of Statistics, Fourth Edition

*Elementary Statistics Using
the Graphing Calculator*, Third Edition

Mario F. Triola

Dutchess Community College

Addison-Wesley
is an imprint of

ISBN-13: 978-0-321-57081-9
ISBN-10: 0-321-57081-2

1 2 3 4 5 6 BB 11 10 09 08

Addison-Wesley
is an imprint of

www.pearsonhighered.com

Preface

This *Minitab Manual,* 11th edition, is a supplement to textbooks in the Triola Statistics Series (listed below), *Biostatistics for the Biological and Health Sciences* by Marc Triola, M.D. and Mario F. Triola, and *Statistical Reasoning for Everyday Life* by Bennett, Briggs, and Triola.

Triola Statistics Series

Elementary Statistics, Eleventh Edition

Elementary Statistics Using Excel, Fourth Edition

Essentials of Statistics, Fourth Edition

Elementary Statistics Using the TI-83/84 Calculator, Third Edition

This manual is based on Minitab Release 16, but it can also be used with other versions of Minitab, including Minitab Release 15 and the *Student Edition of Minitab, R 14*.

The associated textbooks are packaged with a CD-ROM that includes Minitab worksheets. The data sets in Appendix B of the textbook are included as Minitab worksheets on that CD-ROM. Those Minitab worksheets are also available on the web site http://www.aw.com/triola.

Here are major objectives of this manual/workbook and the Minitab software:

- Describe how Minitab can be used for the methods of statistics presented in the textbook. Specific and detailed procedures for using Minitab are included, along with examples of Minitab screen displays.

- Involve students with a spreadsheet type of software application, which will help them work with other spreadsheet programs they will encounter in their professional work.

- Incorporate an important component of computer usage without using valuable class time required for concepts of statistics.

- Replace tedious calculations or manual construction of graphs with computer results.

- Apply alternative methods, such as simulations, that are made possible with computer usage.

- Include topics, such as analysis of variance and multiple regression, that require calculations so complex that they realistically cannot be done without computer software.

It should be emphasized that this manual/workbook is designed to be a *supplement* to textbooks in the Triola Statistics Series; it is not designed to be a self-contained statistics textbook. It is assumed throughout this manual/workbook that the theory, assumptions, and procedures of statistics are described in the textbook that is used.

Chapter 1 of this supplement describes some of the important basics for using Minitab. Chapters 2 through 14 in this manual/workbook correspond to Chapters 2 through 14 in *Elementary Statistics*, 11th edition. However, the individual chapter *sections* in this manual/workbook generally do *not* match the sections in the textbook. Each chapter includes a description of the Minitab procedures relevant to the corresponding chapter in the textbook. This cross-referencing makes it very easy to use this supplement with the textbook.

Chapters include illustrations of Minitab procedures as well as detailed steps describing the use of those procedures. It would be helpful to follow the steps shown in these sections so the basic procedures will become familiar. You can compare your own computer display to the display given in this supplement and then verify that your procedure works correctly. You can then proceed to conduct the experiments that follow.

We welcome any comments or suggestions that you might have for improving this Minitab manual/workbook. Please send them to the Addison–Wesley Statistics Editor.

The author and publisher are very grateful to Minitab, Inc. for the continuing support and cooperation.

M.F.T.
LaGrange, New York
January, 2009

Contents

1

Basics

of Minitab

1-1 Starting Minitab

Minitab Release 16 is the latest version available as of this writing. It can be purchased and downloaded online. As this manual was being written, students could purchase and download Minitab for $99.99, and they could rent Minitab for as little as $29.99 for a semester. In addition to Minitab Release 16 and earlier releases, the *Student Edition of Minitab R 14* can also be used. Although this manual/workbook is based on Minitab Release 16, it can be used for earlier releases of Minitab as well as the *Student Edition of Minitab*.

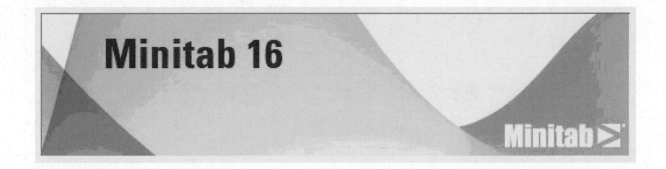

Starting Minitab: With Minitab installed, you can start it by clicking on the Minitab icon on the startup screen.

You can also start Minitab as follows:

1. Click on the taskbar item of **Start**.

2. Select **Programs**.

3. Select **Minitab 16** (or whatever Minitab release you are using).

4. Select **Minitab**.

Session/Data Windows   After opening Minitab, you should see a display of its two main windows: the Session window and the Data window (as shown on the top of the following page). The Session window displays results and it also allows you to enter commands. The top of the display on the following page consists of the Session window. The Data window, shown on the bottom of the display on the following page, consists of a spreadsheet for entering data in columns. These spreadsheets are called *worksheets* in Minitab. Minitab allows you to have multiple worksheets in different Data windows.

Session and Data Windows

Installing Data Sets from the CD-ROM   The CD-ROM included with the textbook includes Minitab worksheets containing data sets in Appendix B from the textbook. It might be helpful to copy those worksheets to the Data folder that is created when Minitab is installed. Although this is not necessary, it could make it easier to access those data sets when they are needed. The Minitab worksheets are in the Data Sets folder on the CD-ROM and, when copying to the Minitab Data folder, answer "yes" when asked if it is OK to overwrite the worksheet BEARS.MTW.

1-2 Entering Data

In Minitab, there are three types of data: (1) *numeric* data (numbers), (2) *text* data (characters), and (3) *date/time* data. We will work mostly with numeric data. Also, data can be in the form of a column, a stored constant, or a matrix. We will work with columns of data. Refer to the next page, which shows a portion of a typical Minitab screen. We have entered data values in the first five columns identified as C1, C2, C3, C4, and C5. Because the entries in the first column consist of text data, the first column becomes C1-T instead of C1. We have also entered the column names of "Time Period," kWh, and so on. (See Data Set 12 in Appendix B of the textbook.)

       Entry of sample data is very easy. Click on the first cell on the desired column, type an entry, then press the **Enter** key. Then type a second entry and press the **Enter** key again. Continue until all of the sample data have been entered. If you make a mistake, simply click on the wrong entry and type the correct value and press **Enter**.

Naming columns: It is wise to enter a name for the different columns of data, so that it becomes easy to keep track of data sets with meaningful names. See the names of "Time Period," kWh, Cost (dollars), and so on.

Worksheets: It often happens that several columns of data are related by being part of one big and happy data set. The following display shows part of a worksheet that contains the measured electricity consumption and cost and heating degree days and average daily temperatures for different time periods. The data are from the author's home. A worksheet can be handled as a single entity that contains one or more columns. A worksheet can be named so that it can be saved and retrieved. See Section 1-3, which follows.

Worksheet

↓	C1-T	C2	C3	C4	C5
	Time Period	kWh	Cost (dollars)	Heating Degree Days	Average Daily Temp
1	Year 1: Jan/Feb	3637	295.33	2226	29
2	Year 1: March/Apr	2888	230.08	1616	37
3	Year 1: May/June	2359	213.43	479	57
4	Year 1: July/Aug	3704	338.16	19	74
5	Year 1: Sept/Oct	3432	299.76	184	66
6	Year 1: Nov/Dec	2446	214.44	1105	47
7	Year 2: Jan/Feb	4463	384.13	2351	28
8	Year 2: March/Apr	2482	295.82	1508	41
9	Year 2: May/June	2762	255.85	657	54

1-3 Saving Data

After entering one or more columns of data, you can *save* all of the data as a Minitab worksheet as follows.

Minitab Procedure for Saving a Worksheet

1. Click on the main menu item of **File**.

2. Click on **Save Current Worksheet As ....**

3. You will see a dialog box, such as the one shown below. In the "Save in" box, enter the location where you want the worksheet saved. In the "File name" box, enter the name of the file. If you omit the extension of .MTW, it will be automatically provided by Minitab.

4. Click on **Save**.

Note: For saved worksheets, Minitab is *forward* compatible, but not *backward* compatible. For example, a worksheet saved in Minitab Release 16 format is not compatible with Minitab Release 15 and earlier releases of Minitab, but a worksheet saved in a Minitab Release 14 format can be opened and used in Minitab 14, Minitab 15, Minitab 16, and later releases.

Saving a Worksheet

1-4   Retrieving Data from the CD-ROM

The CD-ROM packaged with the textbook contains worksheets for the data sets in Appendix B of the textbook. See Appendix B in the textbook for the worksheet names. Here is the procedure for retrieving a worksheet from the CD-ROM:

1.   Click on Minitab's main menu item of **File**.

2.   Click on the subdirectory itcm of **Open Worksheet**.

3.   You should now see a window like the one shown on the following page.

4.   In the "Look in" box at the top, select the location of the stored worksheets. For example, if the CD-ROM is in drive, do this:

- In the "Look in" box, select drive **D** (or whatever drive contains the CD-ROM).

- Double click on the folder **App B Data Sets**.

- Double click on the folder **Minitab**.

- Click on the name of the worksheet that you want to open.

5.   Click on the **Open** button.

The columns of selected data set should now appear in the Minitab display, and they are now available for use.

Opening a Worksheet

1-5 Printing

We have noted that Minitab involves a screen with two separate parts: The Session window at the top and the Data window at the bottom. The contents of the Session window and Data window can be printed as follows.

Printing the Session Window or Data Window

1.   *Session window:* If you want to print the session window portion of the screen, click on the session window.

     Data window: If you want to print the data portion of the screen, click on any cell in the Data window.

2.   Click on the main menu item of **File**.

3.   Click on **Print Session Window** or **Print Worksheet**. (You will see only one of these two options. The option that you see is the result of the choice you made in Step 1. If you see "Print Worksheet" but you want to print the Session window, go back to Step 1 and click on the Session window portion of the screen.)

4.   If you select Print Worksheet in Step 3, you will get a dialog box. Enter your preferences, then click **OK**.

Printing a Large Data Set in a Word Processor

Most of the data sets used with the Triola statistics series of textbooks are not so large that they cannot be printed on a few pages. If you do enter or somehow create a data set with thousands of values, printing would require many pages. One way to circumvent that problem is to move the data to a word processor where the values can be reconfigured for easier printing. Follow these steps.

1.   With the data set displayed in the data window, click on the value at the top. Hold the mouse button down and drag the mouse to the bottom of the data set, then release it so that the entire list of values is highlighted.

2.   Click on **Edit**, then click on **Copy Cells**.

3.   Now go into your word processor and click on **Edit**, then **Paste**. The entire list of values will be in your word processing document where you can configure them as you please. For example, you might press **End**, then press the **space bar**, then press the **Del** key. The second value will be moved up to the top row. Repeat this process to rearrange the data in multiple rows and columns (instead of one really large column). For example, this approach was used to print on *one* page the 175 values in the column labeled CANS109 that is in the Minitab worksheet named CANS. The result is shown below. You can change the number of columns as you desire.

270 273 258 204 254 228 282 278 201 264 265 223 274 230 250 275 281 271 263 277
275 278 260 262 273 274 286 236 290 286 278 283 262 277 295 274 272 265 275 263
251 289 242 284 241 276 200 278 283 269 282 267 282 272 277 261 257 278 295 270
268 286 262 272 268 283 256 206 277 252 265 263 281 268 280 289 283 263 273 209
259 287 269 277 234 282 276 272 257 267 204 270 285 273 269 284 276 286 273 289
263 270 279 206 270 270 268 218 251 252 284 278 277 208 271 208 280 269 270 294
292 289 290 215 284 283 279 275 223 220 281 268 272 268 279 217 259 291 291 281
230 276 225 282 276 289 288 268 242 283 277 285 293 248 278 285 292 282 287 277
266 268 273 270 256 297 280 256 262 268 262 293 290 274 292

1-6 Command Editor and Transforming Data

Data may be *transformed* with operations such as adding a constant, multiplying by a constant, or using functions such as logarithm (common or natural), sine, exponential, or absolute value. For example, if you have a data set consisting of temperatures on the Fahrenheit scale (such as the body temperatures in the Minitab data set BODYTEMP included on the CD) and you want to transform the values to the Celsius scale, you can use the equation

$$C = \frac{5}{9}(F - 32)$$

Such transformations can be accomplished by using Minitab's *command editor* or by using Minitab's virtual calculator.

Using Command Editor to Transform Data

1.   Click on the Session window portion of the screen. (The Session window is on the top portion; the Data window is the bottom portion.)

2.   Click on **Editor**.

3.   Click on **Enable Commands**.

4.   You should now see **MTB >** and you can enter a command. The command
        LET C2 = (5/9)*(C1 — 32)
      tells Minitab to take the values in column C1, subtract 32 from each value, then multiply by the fraction 5/9. This expression corresponds to the above formula for converting Fahrenheit temperatures to the Celsius scale. The temperatures on the Fahrenheit scale of 98.6, 98.6, 98.0, 97.3, and 97.2 are converted to these temperatures on the Celsius scale: 37.0000, 37.0000, 36.6667, 36.2778, and 36.2222.

Using Calc/Calculator to Transform Data

You can also click on **Calc**, then **Calculator** to get a virtual calculator that allows you to create new columns by performing operations with existing columns. The display below shows that column C7 will be the result of converting the temperatures in column C5 from the Fahrenheit scale to the Celsius scale. The expression shown below indicates that 32 is subtracted from each entry in column C5, then the result is multiplied by 5/9. Click **OK** when done.

1-7 Closing Worksheets and Exiting Minitab

Closing worksheets: You can close the current *worksheet* by clicking on the × located at the upper right corner *of the worksheet window*. If you click on that ×, you will be told that the data in the worksheet will be removed and this action cannot be undone. You will be given the opportunity to save the data in a separate file, or you can click **No** to close the window. The Minitab program will continue to be active, even though you have just closed the current worksheet.

Exiting Minitab: Had enough for now? To exit or quit the Minitab program, click on the × located in the extreme upper right corner. Another way to exit Minitab is to click on **File**, then click on **Exit**. In both cases, you will be asked if you want to save changes to the project before closing. This is your last chance to save items that you want saved. Assuming that you are finished and you want to exit Minitab, click **No**.

1-8 Exchanging Data with Other Applications

There may be times when you want to move data from Minitab to another application (such as Excel or STATDISK or Microsoft Word) or to move data from another application to Minitab. Instead of manually retyping all of the data values, you can usually transfer the data set directly. Given below are two ways to accomplish this. The first of the following two procedures is easier, if it works (and it often does work).

Method 1: Use Edit/Copy and Edit/Paste

1.   With the data set displayed in the source program, use the mouse to highlight the values that you want to copy. (Click on the first value, then hold the mouse button down and drag it to highlight all of the values, then release the mouse button.)

2.   While still in the source program, click on **Edit**, then click on **Copy**.

3.   Now go into the software program that is your destination and click on **Edit**, then **Paste**. The entire list of values should reappear.

Method 2: Use Text Files

1.   In the software program containing the original set of data, create a text file of the data. (In Minitab you can create a text file by clicking on **File**, then **Other Files**, then **Export Special Text**.)

2.   Minitab and most other major applications allow you to import the text file that was created. (To import a text file into Minitab, click on **File**, then **Other Files**, then **Import Special Text**.)

CHAPTER 1 EXPERIMENTS: Basics of Minitab

1-1.   ***Entering Sample Data*** When first experimenting with procedures for using Minitab, it's a good strategy to use a small data set instead of one that is large. If a small data set is lost, you can easily enter it a second time. In this experiment, we will enter a small data set, save it, retrieve it, and print it. The Body Temperatures data set from Data Set 2 in Appendix B of the Triola textbook includes these body temperatures, along with others:

> 98.6 98.6 98.0 98.0 99.0 98.4 98.4 98.4 98.4 98.6

    a.   Start Minitab and enter the above sample values in column C1. (See the procedure described in Section 1-2 of this manual/workbook.)

    b.   Save the worksheet using the worksheet name of TEMP (because the values are body temperatures). See the procedure described in Section 1-3 of this manual/ workbook.

    c.   Print the data set by printing the display of the data window. (When first loading Minitab, the data window might not display all 10 values simultaneously, but the data window can be enlarged to display more data. Simply position the mouse on the top border of the data window, then hold the mouse button down to click and drag the top border upward.)

    d.   Exit Minitab, then restart it and retrieve the worksheet named TEMP. Save another copy of the same data set using the file name of TEMP2. Print TEMP2 and include the title at the top.

1-2.   ***Retrieving Data*** The worksheet FHEALTH.MTW (female health) is already stored on the CD-ROM that is included with the textbook. That worksheet contains 13 individual columns of health-related measurements from 40 randomly selected women. Open the worksheet FHEALTH.MTW, then print its contents.

1-3.   ***Generating Random Data*** In addition to entering or retrieving data, Minitab can also *generate* data sets. In this experiment, we will use Minitab to simulate 500 births.

    a. Select **Calc** from the main menu bar, then select **Random Data.**

    b. Select **Integer,** because we want whole numbers.

    c. Enter 500 for the number of values, enter the column (such as C1) to be used for the data, enter 0 for minimum, and enter 1 for the maximum, then click **OK.**

    d. Examine the displayed values and count the number of times that 1 occurs. Record the result here: _____ If we stipulate that a 0 simulates the birth of a boy and 1 simulates the birth of a girl, this experiment simulates 500 births and results in the number of girls.

1-4.   ***Generating Random Data*** Experiment 1-3 results in the random generation of the digits 0 and 1. Minitab can also generate sample data from other types of populations. In this experiment, we will use Minitab to simulate the random selection of 50 IQ scores.

(continued)

a. Select **Calc** from the main menu bar, then select **Random Data.**

b. Select **Normal**. (The normal distribution is described in the textbook.)

c. Enter 50 for the number of values, enter the column (such as C1) to be used for the data, enter 100 for the mean, and enter 15 for the standard deviation. (These statistics are described in the textbook.) Click **OK** and print the results.

1-5. *Saving a Worksheet* Listed below are pulse rates of females (from Data Set 1 in Appendix B in the textbook.) Manually enter the pulse rates in column C1 and name that column Pulse Rate. Save the worksheet with the name Pulse. Print the worksheet.

Female Pulse Rates

76 72 88 60 72 68   80 64 68 68 80 76 68 72 96 72 68 72   64 80
64 80 76 76 76 80 104 88 60 76 72 72 88 80 60 72 88 88 124 64

1-6. *Saving a Worksheet* Listed below are the ages of actresses and actors at the time they won Oscars. (See the Data to Decision project near the end of Chapter 2 in the Triola textbook.) Enter the ages of the winning actresses in column C1 and name that column Actresses. Enter the ages of the winning actors in column C2 and name that column Actors. Save the worksheet with the name OSCAR. Print the worksheet.

Best Actresses

22	37	28	63	32	26	31	27	27	28
30	26	29	24	38	25	29	41	30	35
35	33	29	38	54	24	25	46	41	28
40	39	29	27	31	38	29	25	35	60
43	35	34	34	27	37	42	41	36	32
41	33	31	74	33	50	38	61	21	41
26	80	42	29	33	35	45	49	39	34
26	25	33	35	35	28	30	29	61	

Best Actors

44	41	62	52	41	34	34	52	41	37
38	34	32	40	43	56	41	39	49	57
41	38	42	52	51	35	30	39	41	44
49	35	47	31	47	37	57	42	45	42
44	62	43	42	48	49	56	38	60	30
40	42	36	76	39	53	45	36	62	43
51	32	42	54	52	37	38	32	45	60
46	40	36	47	29	43	37	38	45	

1-7. *Transforming Data* Experiment 1-1 results in saving the worksheet TEMP that contains ten body temperatures in degrees Fahrenheit. Retrieve that data set, then proceed to transform the temperatures to the Celsius scale. Store the Celsius temperatures in column C2. (See Section 1-6 in this manual/workbook.) Print the worksheet containing the original Fahrenheit temperatures in column C1 and the corresponding Celsius temperatures in column C2.

2
Summarizing and Graphing Data

Important note: The topics of this chapter require that you use Minitab to enter data, retrieve data, save files, and print results, as described in Chapter 1 of this manual/ workbook.

Important Characteristics of Data

The textbook states that when describing, exploring, and comparing data sets, the following characteristics are usually extremely important:

    1.   **Center:** Measure of center, which is a representative or average value that gives us an indication of where the middle of the data set is located

    2.   **Variation:** A measure of the amount that the values vary among themselves

    3.   **Distribution:** The nature or shape of the distribution of the data, such as bell-shaped, uniform, or skewed

    4.   **Outliers:** Sample values that are very far away from the vast majority of the other sample values

    5.   **Time:** Changing characteristics of the data over time

In this chapter, we learn how to use Minitab as a tool for investigating the *distribution* of the data by constructing a suitable frequency distribution or graph.

2-1   Frequency Distributions

Section 2-2 in the textbook describes the construction of a table representing the *frequency distribution* for a set of data. Shown below are pulse rates of females, followed by Table 2-2 from the textbook. Table 2-2 is a frequency distribution summarizing the female pulse rates. If Exercise 5 in Chapter 1 of this manual/workbook was completed, these data are saved as a worksheet named Pulse.

Female Pulse Rates
76 72 88 60 72 68 80 64 68 68 80 76 68 72 96 72 68 72 64 80
64 80 76 76 76 80 104 88 60 76 72 72 88 80 60 72 88 88 124 64

Table 2-2   Pulse Rates of Females

Pulse Rate	Frequency
60-69	12
70-79	14
80-89	11
90-99	1
100-109	1
110-119	0
120-129	1

Minitab does not have a specific feature for generating a frequency distribution as in Table 2-2, but the following procedure can be very helpful in constructing such a table.

1.   Enter the data in column C1 of the Minitab worksheet.

2.   Click on **Statistics,** then click on **Tables**.

3.   Select **Tally Individual Variables**.

4.   Enter **C1** (or the column name) in the box labeled "Variables."

5.   In the pop-up window, select the option of **Counts**.

6.   Click **OK**.

If the preceding pulse rates are used, the result will be as shown below.

Tally for Discrete Variables: C1

C1	Count
60	3
64	4
68	5
72	8
76	6
80	6
88	5
96	1
104	1
124	1
N=	40

With these tally counts, it is much easier to construct the frequency distribution in the format of the preceding Table 2-2. For example, it is easy to see from this tally list that there are 12 data values falling in the first class of 60−69.

Another way to obtain the frequency distribution is to generate a stemplot as described in Section 2-4 of this manual/workbook. The frequency counts needed for the frequency distribution can be identified by simply counting the number of digits to the right of each stem.

2-2 Histograms

Textbooks in the Triola textbook series describe histograms and provide detailed procedures for constructing them. It is noted that a histogram is an excellent device for exploring the *distribution* of a data set. Here is the Minitab procedure for generating a histogram.

1.   Enter the data in a Minitab column, such as column C1. If the data are already stored in a Minitab worksheet, retrieve the worksheet using the procedure described in Section 1-4 of this manual/workbook.

2.   Click on the main menu item of **Graph**.

3.   Select **Histogram** from the subdirectory.

4.   Click on the option of **Simple**, then click **OK**.

5.   In the dialog box, click on the column with the data so that the column label (such as C1) or the data name (Pulse Rate) appears in the box as shown.

Generating a Histogram

6.   Using Minitab with the preceding pulse rates results in the following histogram. Figure 2-3 in the textbook shows another histogram with different class boundaries. The following Minitab-generated histogram can be edited by double-clicking on the attribute that you want to change. To change the horizontal scale to correspond to Figure 2-3 in the textbook, double-click on the *x*-axis and use the pop-up window to change the scale. For example, in the Edit Scale pop-up window, click on the **Binning** tab, select **Midpoint**, select **Midpoint/Cutpoint positions**, and enter the class midpoint values of 65, 75, 85, 95, 105, 115, 125. Click **OK** and the second graph on the following page will be displayed. This second graph corresponds to Figure 2-3 in the textbook.

Default Histogram

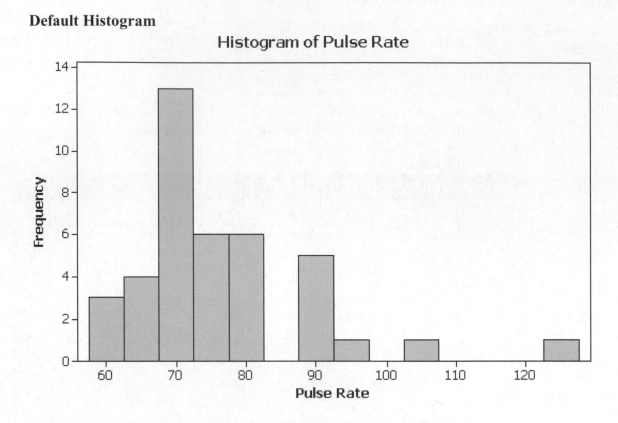

Histogram with Class Midpoints of 65, 75, 85, 95, 105, 115, 125

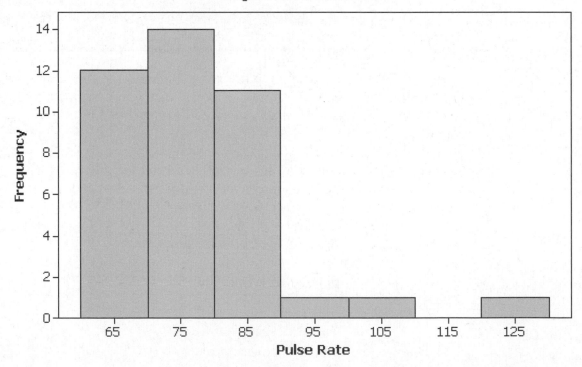

2-3 Dotplots

Here is the Minitab procedure for creating a dotplot.

1. Enter or retrieve the data into a Minitab column.

2. Click on the main menu item of **Graph**.

3. Select **Dotplot**.

4. Select the **Simple** dotplot and click **OK**.

5. The dialog box will list the columns of data at the left. Click on the desired column. You can edit a dotplot or any other graph by double-clicking on the attribute you want to modify. To change the scale, double-click on the axis and use the pop-up window.

6. Click **OK**.

Shown below is the Minitab dotplot display of the pulse rates of females listed in Section 2-1 of this manual/workbook. The original dotplot showed values of 63, 72, 81, 90, 99, 108, 117 along the horizontal scale, but the dotplot was modified to show the more convenient values of 60, 65, 70, 75, 80, 85, 90, 95, 100, 105, 110, 115, 120, 125 instead. This change in the numbers used for the horizontal scale was made by double-clicking on a number on the horizontal scale, selecting the **scale** tab, then manually entering the values of 60, 65, 70, 75, 80, 85, 90, 95, 100, 105, 110, 115, 120, 125 in the box labeled **Position of ticks**.

2-4 Stemplots

Minitab can provide stemplots (or stem-and-leaf plots). The textbook notes that a **stemplot** represents data by separating each value into two parts: the stem (such as the leftmost digit) and the leaf (such as the rightmost digit). Here is the Minitab procedure for generating a stemplot.

1. Enter or retrieve the data into a Minitab column.

2. Click on the main menu item of **Graph**.

3. Select **Stem-and-Leaf**.

4. The dialog box lists columns of data at the left. Click on the desired column.

5. Remove the check mark for "Trim outliers," so that outliers will be included.

6. Enter a desired value for the class interval, such as 10.

7. Click **OK**.

Shown below is the Minitab stemplot based on the pulse rates of females listed in Section 2-1 of this manual/workbook. The actual stem–and–leaf plot is highlighted below with a bold font. This stemplot is based on the use of 10 as the increment value.

Stem-and-Leaf Display: Pulse Rate

```
Stem-and-leaf of Pulse Rate   N  = 40
Leaf Unit = 1.0

  12    6    000444488888
 (14)   7    22222222666666
  14    8    00000088888
   3    9    6
   2   10    4
   1   11
   1   12    4
```

Reading and Interpreting a Minitab Stemplot: In addition to the stemplot itself, Minitab also provides another column of data at the extreme left as shown above. In the above display, *the leftmost column represents cumulative totals.* The left column above shows that there are 12 sample values included between 60 and 69. The left column entry of (14) indicates that there are 14 data values in the row containing the median. The left-column entries *below* the median row represent cumulative totals from the bottom up, so that the 1 at the bottom indicates that there is one value between 120 and 129. This leftmost column of values is tricky to interpret, so it might be a good idea to just ignore it.

2-5 Bar Graphs

Shown below is a Minitab-generated bar graph similar to Figure 2-1 in the textbook. The procedure for generating such a bar graph is as follows.

1. Enter the *names* of the categories (such as Democrats, Republicans, Independents) in column C1.

2. Enter the corresponding frequencies or percentage values in column C2.

3. Click on **Graph**.

4. Click on **Bar Chart.**

5. In the "Bars represent" box at the top, select the option of **Values from a table**.

6. Click on the **Simple** bar graph at the left, then click **OK**.

7. In the "Graph variables" box, enter C2 (or whatever column contains the frequencies or percentage values).

8. In the "Categorical variable" box, enter C1 (or whatever column contains the names of the categories).

9. Click **OK**.

Bar Graph

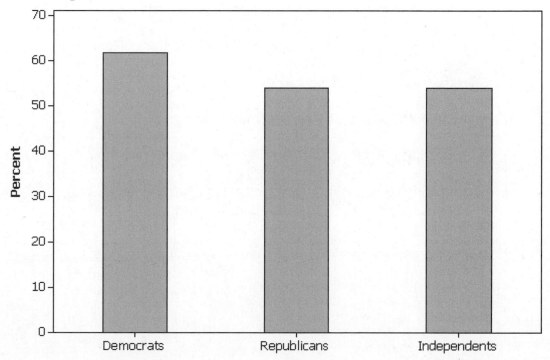

The textbook notes that a **multiple bar graph** has two or more sets of bars and is used to compare two or more data sets. Section 2-4 in the textbook includes an example of a multiple bar graph. The above procedure can be used for a multiple bar graph, but use the **Cluster** option in Step 6, not the option of Simple.

2-6 Pareto Charts

Section 2-4 of the textbook includes the Minitab-generated Pareto chart shown below. A Pareto chart is a bar graph for categorical data, with the bars arranged in order according to frequencies. The tallest bar is at the left, and the smaller bars are farther to the right, as shown below.

Pareto Chart

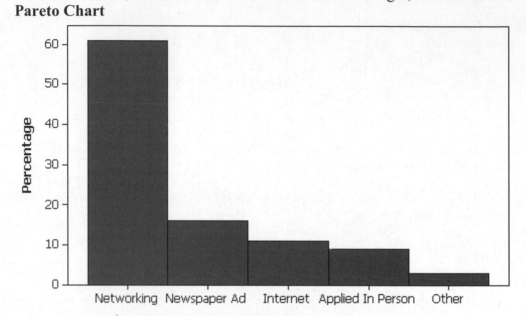

Here is the procedure for using Minitab to construct a Pareto chart:

1.   Enter the *names* of the categories in column C1.

2.   Enter the corresponding frequency counts in column C2.

3.   Select the menu items of **Stat**, **Quality Tools**, and **Pareto Chart**.

4.   When the dialog box is displayed, select the **Charts defect table** option and enter C1 in the labels box and C2 in the frequency box.

5.   Click **OK**.

2-7 Pie Charts

To obtain a pie chart from Minitab, follow this procedure.

1. Enter the *names* of the categories in column C1.

2. Enter the corresponding frequency counts in column C2.

3. Select the main menu item of **Graph**.

4. Select **Pie Chart**.

5. When the dialog box is displayed, select **Chart values from a table** and enter C1 for the categorical variable box and enter C2 in the summary variables box.

6. Click **OK**. Shown below is the Minitab pie chart included in Section 2-4 of the textbook.

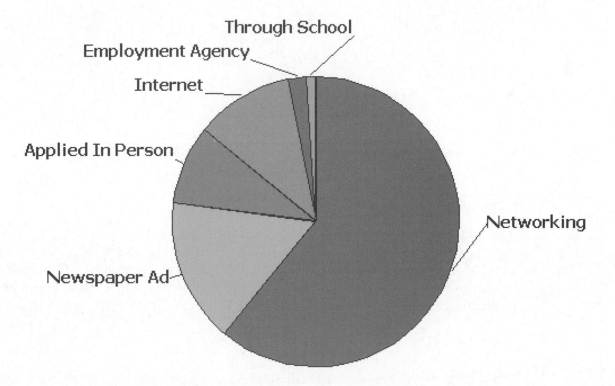

2-8 Scatterplots

The textbook describes a scatterplot (or scatter diagram). A scatterplot can be very helpful in seeing a relationship between two variables. The scatterplot shown below results from paired data consisting of the number of chirps per minute by a cricket and the temperature. This scatterplot shows that as the number of chirps increases, the corresponding temperature tends to be higher.

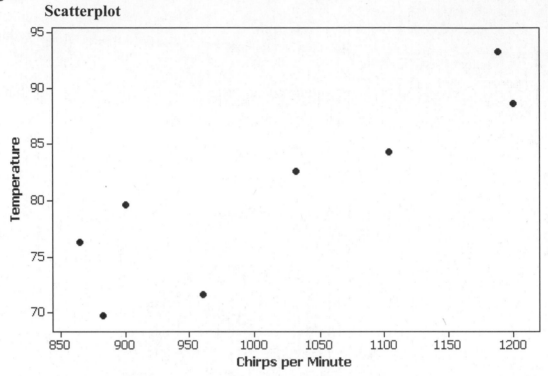

Here is the Minitab procedure for generating a scatterplot.

1. Given a collection of paired data, enter the values for one of the variables in column C1, and enter the corresponding values for the second variable in column C2. (You can also use any two Minitab columns other than C1 and C2.) Be careful to enter the two columns of data so that they are matched in the same way that they are paired in the original listing.

2. Click on **Graph** from the main menu.

3. Select the subdirectory item of **Scatterplot.** Select the option of **Simple**, then click **OK**.

4. In the dialog box, enter C1 in the box for the Y variable, then enter C2 in the box for the X variable (or vice versa).

5. Click **OK**.

Another procedure for generating a scatterplot is to select **Stat**, then **Regression**, then **Fitted Line Plot**. We will use this option later when we discuss the topics of correlation and regression. For now, we are simply generating the scatter diagram to visually explore whether there is an obvious pattern that might reveal some relationship between the two variables.

2-9 Time–Series Graph

Time–series data are data that have been collected at different points in time, and Minitab can graph such data so that patterns become easier to recognize. Here is the Minitab procedure for generating a time–series graph.

1. Given a collection of time–series data, enter the sequential values in column C1.

2. Select the main menu item of **Graph**.

3. Select the menu item of **Time Series Plot.**

4. Select the option of **Simple**.

5. A dialog box now appears. Enter C1 in the first cell under Y in the "Graph" list or, if using Minitab Release 14, enter C1 in the "Series" box.

6. Enter a start time to be used for the horizontal axis. Click the **Options** button to enter a start time; click **Time/Scale** and proceed to select the desired time scale (such as "Calendar") along with a start value (such as 1980) and an increment value (such as 1). Click **OK** twice.

Listed below are the high values of the Dow Jones Industrial Average (DJIA) for each year from 1980 to 2006. The Minitab time-series graph follows. When compared to the list of frequency values, we can see that the time-series graph does a much better job of showing the behavior of that stock market index.

1000 1024 1071 1287 1287 1553 1956 2722 2184 2791 3000 3169 3413 3794
3978 5216 6561 8259 9374 11568 11401 11350 10635 10454 10855 10941 12464

Time-Series Graph of DJIA High Values

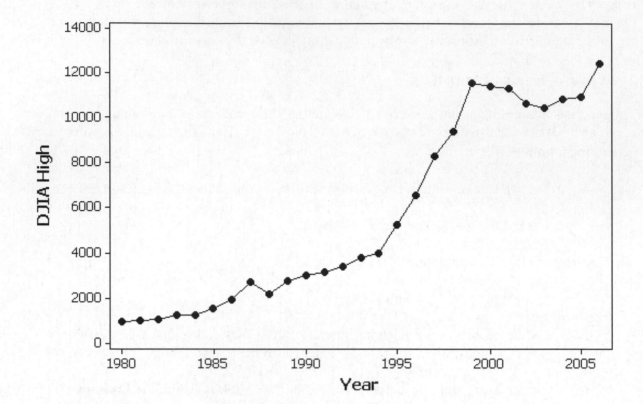

This chapter has described procedures for using Minitab to construct frequency distributions and a variety of different graphs. Boxplots are also helpful graphs, and they will be described in Chapter 3 of this manual/workbook.

CHAPTER 2 EXPERIMENTS: **Summarizing and Graphing Data**

In Experiments 2-1 through 2-8, use the data with Minitab to construct a histogram. Each experiment is from the textbook, but the number of classes and the starting point is not specified.

2-1.   *Analysis of Last Digits* Heights of statistics students were obtained by the author as part of an experiment conducted for class. The last digits of those heights are listed below.

　　　　　0 0 0 0 0 0 0 0 1 1 2 3 3 3 4 5 5 5
　　　　　5 5 5 5 5 5 5 5 5 5 5 5 5 6 6 8 8 8 9

2-2.   *Radiation in Baby Teeth* Listed below are amounts of Strontium-90 (in millibecquerels) in a simple random sample of baby teeth obtained from Pennsylvania residents born after 1979 (based on data from "An Unexpected Rise in Strontium-90 in U.S. Deciduous Teeth in the 1990s," by Mangano, et. al., *Science of the Total Environment*).

155 142 149 130 151 163 151 142 156 133 138 161 128 144 172 137 151 166
147 163　145 116 136 158 114 165 169 145 150 150 150 158 151 145 152 140
170 129 188 156

2-3.   *Nicotine in Non-Filtered Cigarettes* Refer to Data Set 4 in Appendix B and use the 25 nicotine amounts (in mg) listed for the non-filtered king-size cigarettes.

2-4.   *Nicotine in Filtered Cigarettes* Refer to Data Set 4 in Appendix B and use the 25 nicotine amounts (in mg) listed for the filtered and non-menthol cigarettes.

2-5.   *Home Voltage Measurements* Refer to Data Set 13 in Appendix B and use the 40 home voltage measurements. Does the result appear to have a normal distribution? Why or why not?

2-6.   *Generator Voltage Measurements* Refer to Data Set 13 in Appendix B and use the 40 voltage measurements from the generator. Using a very loose interpretation of the relevant criteria, does the result appear to have a normal distribution? Compare the frequency distribution to the result from the preceding exercise.

2-7.   *How Long is a 3/4 in. Screw?* Refer to Data Set 19 in Appendix B and use the 50 screw lengths to construct a frequency distribution. The screws were labeled as having a length of 3/4 in. Does the frequency distribution appear to be consistent with the label? Why or why not?

2-8.   *Weights of Discarded Paper* As part of the Garbage Project at the University of Arizona, the discarded garbage was analyzed for 62 households. Use the 62 weights of discarded paper from Data Set 22 in Appendix. Do the weights of discarded paper appear to have a normal distribution?

In Experiments 2-9 and 2-10, refer to the data and use Minitab to construct a dotplot.

2-9.   ***Dotplot*** Use the data from Experiment 2-1.

2-10.   ***Dotplot*** Use the data from Experiment 2-2.

In Experiments 2-11 and 2-12, refer to the data and use Minitab to construct a stemplot.

2-11.   ***Stemplot*** Use the data from Experiment 2-1.

2-12.   ***Stemplot*** Use the data from Experiment 2-2.

2-13.   ***Pareto Chart*** The following table (based on data from the U.S. National Center for Education Statistics) shows the distribution of undergraduate college student enrollments. Use Minitab to construct a Pareto chart for the data in this table.

College	Relative Frequency
Public 2-Year	36.8%
Public 4-Year	40.0%
Private 2-Year	1.6%
Private 4-Year	21.9%

2-14.   ***Pie Chart*** Use Minitab to construct a pie chart based on the data in Experiment 2-13.

2-15.   ***Pie Chart*** Chief financial officers of U. S. companies were surveyed about areas in which job applicants make mistakes. Here are the areas and the frequency of responses: interview (452); resume (297); cover letter (141); reference checks (143); interview follow-up (113); screening call (85). These results are based on data from Robert Half Finance and Accounting. Use Minitab to construct a pie chart representing the given data.

2-16.   ***Pareto Chart*** Use Minitab to construct a Pareto chart of the data given in Experiment 2-15. Compare the Pareto chart to the pie chart. Which graph is more effective in showing the relative importance of the mistakes made by job applicants?

In Experiments 2-17 and 2-18, use Minitab to construct a scatterplot.

2-17.   ***Scatterplot*** In Data Set 3 from Appendix B in the textbook, use tar in king-size cigarettes for the horizontal scale and use carbon monoxide (CO) in king-size cigarettes for the vertical scale. Determine whether there appears to be a relationship between cigarette tar and CO in king-size cigarettes. If so, describe the relationship.

2-18. ***Scatterplot*** In Data Set 12 from Appendix B, use the 22 average daily temperatures and use the corresponding 22 amounts of energy consumption (kWh). (Use the temperatures for the horizontal scale.) Based on the result, is there a relationship between the average daily temperatures and the amounts of energy consumed? Try to identify at least one reason why there is (or is not) a relationship.

2-19. *Time-Series Graph* In 1965, Intel co-founder Gordon Moore initiated what has since become known as *Moore's law:* the number of transistors per square inch on integrated circuits will double approximately every 18 months. The table below lists the number of transistors per square inch (in thousands) for several different years. The top row lists years and the bottom row lists the numbers of transistors. Use Minitab to construct a time-series graph of the data.

1971	1974	1978	1982	1985	1989	1993	1997	1999	2000	2002	2003
2.3	5	29	120	275	1180	3100	7500	24,000	42,000	220,000	410,000

2-20. *Time-Series Graph* The following table shows the numbers of cell phone subscriptions (in thousands) in the United States for various years. (The years are listed in the top row.) Use Minitab to construct a time-series graph of the data. "Linear" growth would result in a graph that is approximately a straight line. Does the time-series graph appear to show linear growth?

1985	1987	1989	1991	1993	1995	1997	1999	2001	2003	2005
340	1231	3509	7557	16,009	33,786	55,312	86,047	128,375	158,722	207,900

2–21. ***Working with Your Own Data*** Through observation or experimentation, collect your own set of sample values. Obtain at least 40 values and try to select data from an interesting population. Use Minitab to generate graphs suitable for describing the distribution of the data. Describe the data and important characteristics.

_____

_____

_____

_____

_____

_____

3

Statistics for Describing, Exploring, and Comparing Data

Important note: The topics of this chapter require that you use Minitab to enter data, retrieve data, save files, and print results. These functions are described in Chapter 1 of this manual/ workbook. Be sure to understand those functions from Chapter 1 before beginning this chapter.

The textbook notes that these characteristics of a data set are extremely important: center, variation, distribution, outliers, and changing characteristics of data over time. Chapter 2 of this manual/workbook presented a variety of graphs that are helpful in learning about the distribution of a data set. This chapter presents methods for learning about center, variation, and outliers.

3-1 Descriptive Statistics

Let's consider the following lists of pulse rates of females and males. These values are listed in the Minitab worksheets FHEALTH (female health measurements) and MHEALTH (male health measurements) that are on the CD included with the textbook.

Female Pulse Rates
76 72 88 60 72 68   80 64 68 68 80 76 68 72 96 72 68 72   64 80
64 80 76 76 76 80 104 88 60 76 72 72 88 80 60 72 88 88 124 64

Male Pulse Rates
68 64 88 72 64 72 60 88 76 60 96 72 56 64 60 64 84 76 84 88
72 56 68 64 60 68 60 60 56 84 72 84 88 56 64 56 56 60 64 72

Given a data set, such as the above list of female pulse rates, we can use Minitab to obtain descriptive statistics, including the mean, standard deviation, and quartiles. Here is the Minitab procedure.

1. Enter or retrieve the data in a Minitab column, such as C1. (See Section 1-2.)

2. Click on the main menu item of **Stat**.

3. Click on the subdirectory item of **Basic Statistics**.

4. Select **Display Descriptive Statistics**.

5. A "Display Descriptive Statistics" dialog box will pop up. In the "Variables" box enter the column containing the data that you are investigating. You can manually enter the column, such as C1, or the name of the column, or you can click on the desired column displayed at the left.
    - You can also click on the **Graphs** bar to generate certain graphs, including a histogram and boxplot.
    - You can also click on the **Statistics** bar to select the particular statistics that you want included among the results.

6. Click **OK**.

If you use the pulse rates of females listed on the preceding page and follow the above procedure, you will get a Minitab display like the one shown below. For this display, the author clicked on the **Statistics** bar and selected the statistics that are particularly relevant to discussions in the textbook.

Descriptive Statistics

```
          Total
Variable  Count   Mean   StDev  Variance  Minimum      Q1
PULSE        40   76.30  12.50    156.22    60.00   68.00


Median     Q3  Maximum    Range
74.00    80.00   124.00    64.00
```

From the above Minitab display, we obtain the following important descriptive statistics (expressed by applying the round-off rule of using one more decimal place than in the original data values).

- Number of data values: 40
- Mean: $\bar{x} = 76.3$
- Standard deviation: $s = 12.5$
- Variance: $s^2 = 156.2$
- Range: 64.0
- Five-number-summary:
  - Minimum: 60
  - First quartile Q_1: 68.0
  - Median: 74.0
  - Third quartile Q_3: 80.0
  - Maximum: 124

Quartiles: The textbook uses a simplified procedure for finding quartiles, so the quartiles Q_1 and Q_3 found with Minitab may differ slightly from those found by using the procedure described in the textbook. For the above pulse rates of females, Minitab's quartiles are the same as that would be obtained by using the method described in the textbook, but Minitab's quartiles will not always be exactly the same as those found from the procedure in the textbook.

3-2 z Scores

The textbook describes z scores (or standard scores). For sample data with mean $\bar{x}$ and standard deviation s, the z score can be found for a sample value x by computing

$$z = \frac{x - \bar{x}}{s} \quad \text{or} \quad z = \frac{x - \mu}{\sigma}$$

Here is the Minitab procedure for finding z scores corresponding to sample values. (This procedure uses column C1, but any other Minitab column can be used instead.)

1. Enter a list of data values in column C1.

2. Click on the main menu item of **Calc**.

3. Select **Standardize** from the subdirectory.

4. A dialog box should appear. Enter C1 for the input column and C2 (or any other specific column) for the column in which the results will be stored.

5. There are buttons for different calculations, but be sure to select the button for "Subtract mean and divide by standard deviation."

6. Click **OK**, and column C2 will magically appear with the z score equivalent of each of the original sample values. Because column C2 is now available as a set of data, you can explore it using Descriptive Statistics, Histogram, Dotplot, and so on.

Shown below are the first few rows from the Minitab display of the z scores for the above list of pulse rates of females. The column label of "z score" was manually entered.

Pulse Rate	z score
76	-0.02400
72	-0.34404
88	0.93610
60	-1.30414
72	-0.34404
68	-0.66407

3-3 Boxplots

The textbook describes the construction of boxplots based on the minimum value, maximum value, median, and quartiles Q_1 and Q_3. *Note:* The values of Q_1 and Q_3 generated by Minitab may be slightly different from the values obtained by using the procedure described in the textbook. Here is the Minitab procedure for generating boxplots.

1.   Enter or retrieve the data into a column of Minitab.

2.   Click on the main menu item of **Graph**.

3.   Select the subdirectory item of **Boxplot**.

4.   Select **One Y Simple** for one boxplot or select **Multiple Y's Simple** for boxplots of two or more data sets.

5.   In the dialog box, click on the column(s) containing the data for which the boxplot will be produced.

6.   Click **OK**.

Shown below are the two boxplots for the pulse rates of females and males listed near the beginning of this section. Boxplots are particularly useful for comparing data sets. The asterisks are used to identify data values that appear to be outliers. By comparing the two boxplots, we see that the pulse rates of females appear to be somewhat higher than those of males.

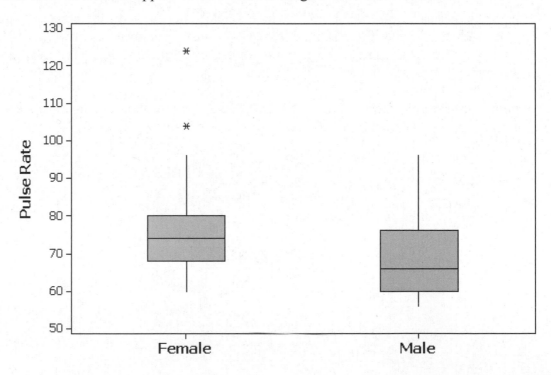

3-4 Sorting Data

There is often a need to *sort* data. Data are sorted when they are arranged in order from low to high (or from high to low). Identification of outliers becomes easier with a sorted list of values. Here is how we can use Minitab to sort data.

1.   Enter or retrieve a data set into a column of Minitab.

2.   Click on the main menu item of **Data.**

3.   Click on the subdirectory item of **Sort**.

4.   You will now see a dialog box like the one shown below. In the "Sort column(s)" box, enter the column that you want to sort. In the "By column" box, enter the column to be used as the basis for sorting (this is usually the column containing the original unsorted data). In the "Column(s) of current worksheet" box, enter the column where you want the sorted data to be placed. (You can place the sorted data in the same column as the original data.) The entries shown below tell Minitab to sort the pulse rate data (in column C5) and store the sorted list in column C15. Click **OK** after completing the dialog box.

3-5 Outliers

Minitab does not have a specific function for identifying outliers, but some of the functions described in this chapter can be used. One easy procedure for identifying outliers is to *sort* the data as described in Section 3-4 of this manual/workbook, then simply examine the values near the beginning of the list or the end of the list. Consider a value to be an outlier if it is very far away from almost all of the other data values.

Another way to identify outliers is to construct a boxplot. We have noted that Minitab uses asterisks to depict outliers that are identified with a specific criterion that uses the *interquartile range* (IQR). The interquartile range IQR is as follows:

$$IQR = Q_3 - Q_1$$

Minitab identifies data values as outliers and denotes them with asterisks in a boxplot by applying these same criteria given in Part 2 of Section 3-4 from the textbook:

A data value is an outlier if it is …

> **above Q_3 by an amount greater than $1.5 \times$ IQR**
>
> *or*   **below Q_1 by an amount greater than $1.5 \times$ IQR.**

See the boxplots given in Section 3-3 of this manual/workbook and note that the pulse rates of females are depicted with a boxplot that includes two asterisks corresponding to the two highest pulse rates of 104 and 124. When analyzing these pulse rates, it is a good strategy to consider the effects of those two values by considering results with those values included and the results obtained when those two values are excluded.

Remember that important characteristics of data sets are center, variation, distribution, outliers, and time (that is, changing characteristics of data over time). Outliers are important for a variety of reasons. Some outliers are errors that should be corrected. Some outliers are correct values that may have a very dramatic effect on results and conclusions, so it is important to know when outliers are present so that their effects can be considered.

CHAPTER 3 EXPERIMENTS:
Statistics for Describing, Exploring, and Comparing Data

3–1.   ***Comparing Heights of Men and Women*** In this experiment we use two small data sets as a quick introduction to using some of the basic Minitab features. (When beginning work with new software, it is wise to first work with small data sets so that they can be entered quickly if they are lost or damaged.) The data listed below are measured heights (in inches) of random samples of men and women (taken from Data Set 1 in Appendix B of the textbook).

Men	70.8	66.2	71.7	68.7	67.6	69.2
Women	64.3	66.4	62.3	62.3	59.6	63.6

a. Find the indicated characteristics of the heights of *men* and enter the results below.

Center: Mean: _____   Median: _____

Variation: St. Dev.:_____   Range: _____

5-Number Summary: Min.:_____ Q_1:_____ Q_2:_____ Q_3:_____ Max.:_____

Outliers: _____

b. Find the characteristics of the heights of *women* and enter the results below.

Center: Mean: _____   Median: _____

Variation: St. Dev.:_____   Range: _____

5-Number Summary: Min.:_____ Q_1:_____ Q_2:_____ Q_3:_____ Max.:_____

Outliers: _____

c. Compare the results from parts a and b.

_____

_____

_____

3–2. ***Working with Larger Data Sets*** Repeat Experiment 3–1, but use the sample data for all 40 males and 40 females included in Data Set 1 that is found in Appendix B of the textbook. Instead of manually entering the 80 individual heights (which would be no fun at all), open the worksheets MHEALTH and FHEALTH that are found among the Minitab data sets on the CD–ROM that is included with the textbook.

a. Find the indicated characteristics of the heights of *men* and enter the results below.

Center: Mean: _____   Median: _____

Variation: St. Dev.:_____   Range: _____

5-Number Summary: Min.:_____ Q_1:_____ Q_2:_____ Q_3:_____ Max.:_____

Outliers: _____

b. Find the characteristics of the heights of *women* and enter the results below.

Center: Mean: _____   Median: _____

Variation: St. Dev.:_____   Range: _____

5-Number Summary: Min.:_____ Q_1:_____ Q_2:_____ Q_3:_____ Max.:_____

Outliers: _____

c. Compare the results from parts a and b.

_____

_____

_____

d. Are there are notable differences observed from the complete sets of sample data that could not be seen with the smaller samples listed in Experiment 3–1? If so, what are they?

_____

_____

_____

3–3.   ***Boxplots*** Use the same sets of data used in Experiment 3–2 and print boxplots for the heights of the 40 men and the heights of the 40 women. Include both boxplots in the same window so that they can be compared. Do the boxplots suggest any notable differences in the two sets of sample data?

_____

_____

Interpret the *asterisk* that appears in the Minitab display for the boxplot depicting the heights of the men.

_____

_____

3–4.   ***Effect of Outlier*** In this experiment we will study the effect of an *outlier*. Use the same heights of *men* used in Experiment 3-1, but change the first entry from 70.8 in. to 708 in. (This type of mistake often occurs when the key for the decimal point is not pressed with enough force.) The outlier of 708 in. is clearly a mistake, because a male with of height of 708 in. would be 59 feet tall, or about six stories tall. Although this outlier is a mistake, outliers are sometimes correct values that differ substantially from the other sample values.

Men: 708 66.2 71.7 68.7 67.6 69.2

Using this modified data set with the height of 70.8 in. changed to be the outlier of 708 in., find the following.

   Center: Mean: _____   Median: _____

   Variation: St. Dev.:_____   Range: _____

   5-Number Summary: Min.:_____ Q_1:_____ Q_2:_____ Q_3:_____ Max.:_____

   Outliers: _____

Based on a comparison of these results to those found in Experiment 3–1, how is the mean affected by the presence of an outlier?

_____

How is the median affected by the presence of an outlier?

_____

How is the standard deviation affected by the presence of an outlier?

_____

3-5.   ***Describing Data*** Use the 40 BMI (body mass index) indices of women from the data Set named Female Health Exam Results and enter the results indicated below.

Center: Mean: _____   Median: _____

Variation: St. Dev.:_____   Range: _____

5-Number Summary: Min.:_____ Q_1:_____ Q_2:_____ Q_3:_____   Max.:_____

Outliers: _____

3-6.   ***Describing Data*** Use the 40 BMI (body mass index) indices of men from the data set named Male Health Exam Results and enter the results indicated below.

Center: Mean: _____   Median: _____

Variation: St. Dev.:_____   Range: _____

5-Number Summary: Min.:_____ Q_1:_____ Q_2:_____ Q_3:_____   Max.:_____

Outliers: _____

Compare the BMI indices of women (from Experiment 3-5) and men.

_____

_____

3-7.   ***Boxplots*** Use the BMI indices of females (in data set named Female Health Exam Results and use the BMI indices of males (in data set named Male Health Exam Results). Print their two boxplots together in the same display. (*Hint:* Because the two lists are in different data sets, use Copy/Paste to configure the two lists of values so that they are in the same Sample Editor window.) Do the boxplots suggest any notable differences in the two sets of sample data?

_____

_____

3-8.   ***Sorting Data*** One of the larger data sets in Appendix B in the textbook is the list of NASA shuttle flight durations (in hours), which is included in the Minitab worksheet NASA stored on the CD-ROM provided with the textbook. *Sort* that data set by arranging the shuttle flight durations from lowest to highest. List the first ten values in the sorted column.

_____

3-9.   *z Scores*   Retrieve the Minitab worksheet CANS.MTW, then use Minitab to find the z score corresponding to each of the 175 values for the cans that are 0.0109 in. thick. [Each value is an axial load, which is the weight (in pounds) that the can supports before being crushed.] Store the 175 z scores in column C3, then find the following.

    Center: Mean: _____   Median: _____

    Variation: St. Dev.:_____   Range: _____

    5-Number Summary: Min.:_____ Q_1:_____ Q_2:_____ Q_3:_____ Max.:_____

    Outliers: _____

What is notable about the above results? Specifically, what is notable about the value of the mean and standard deviation?

_____

Will the same mean and standard deviation be obtained for *any* set of sample data? Explain how you arrived at your answer.

_____

_____

3-10.   ***Combining Data***   Open the Minitab worksheet M&M, which consists of six different columns of data. Use **Data/Stack** to make a copy of the data all stacked together in column C7 of the current worksheet.

a. Find the following results for the combined data in column C7.

    Center: Mean: _____   Median: _____

    Variation: St. Dev.:_____   Range: _____

    5-Number Summary: Min.:_____ Q_1:_____ Q_2:_____ Q_3:_____ Max.:_____

    Outliers: _____

b. Print a boxplot of the data set.

c. Describe the important characteristics of the data set. Be sure to address the nature of the distribution, measures of center, measures of variation, and any other important and notable features.

_____

_____

3–11. ***Statistics from a Dotplot***   Shown below is a Minitab dotplot. Identify the values represented in this graph, enter them in column C1, then find the indicated results.

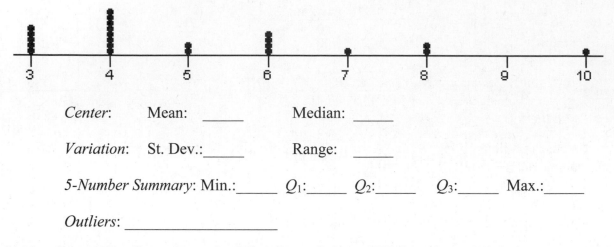

       Center: Mean: _____   Median: _____

       Variation: St. Dev.:_____   Range: _____

       5-Number Summary: Min.:_____ Q_1:_____ Q_2:_____ Q_3:_____ Max.:_____

       Outliers: _____

3–12. ***Comparing Data***   Open the Minitab worksheet COLA and use Minitab to compare the weights of regular Coke and the weights of diet Coke. Obtain printouts of relevant results. What do you conclude? Can you explain any substantial difference?

_____

_____

3–13. ***Comparing Data***   In "Tobacco and Alcohol Use in G–Rated Children's Animated Films," by Goldstein, Sobel and Newman (*Journal of the American Medical Association,* Vol. 281, No. 12), the lengths (in seconds) of scenes showing tobacco use and alcohol use were recorded for animated children's movies. Refer to the data in Appendix B from the textbook and find the mean and median for the tobacco times, then find the mean and median for the alcohol times. Does there appear to be a difference between those times? Which appears to be the larger problem: scenes showing tobacco use or scenes showing alcohol use?

_____

_____

3–14. ***Working with Your Own Data***   Through observation or experimentation, collect your own set of sample values. Obtain at least 40 values and try to select data from an interesting population. List the values below.

_____

_____

_____

_____

Find the following results.

   Center: Mean: _____   Median: _____

   Variation: St. Dev.:_____   Range: _____

   5-Number Summary: Min.:_____ Q_1:_____ Q_2:_____ Q_3:_____ Max.:_____

   Outliers: _____

Use Minitab to further explore the data. Obtain printouts of relevant results. Describe the nature of the data. That is, what do the values represent? Describe important characteristics of the data set, and include Minitab printouts to support your observations.

4

Probabilities Through Simulations

4-1 Simulation Methods

Chapter 4 in the Triola textbook focuses on principles of probability theory. That chapter presents a variety of rules and methods for finding probabilities of different events. The textbook focuses on traditional approaches to computing probability values, but this chapter in this manual/workbook focuses instead on an alternative approach based on *simulations*.

> A **simulation** of a procedure is a process that behaves the same way as the procedure, so that similar results are produced.

Mathematician Stanislaw Ulam once studied the problem of finding the probability of winning a game of solitaire, but the theoretical computations involved were too complicated. Instead, Ulam took the approach of programming a computer to simulate or "play" solitaire hundreds of times. The ratio of wins to total games played is the approximate probability he sought. This same type of reasoning was used to solve important problems that arose during World War II. There was a need to determine how far neutrons would penetrate different materials, and the method of solution required that the computer make various random selections in much the same way that it can randomly select the outcome of the rolling of a pair of dice. This neutron diffusion project was named the Monte Carlo Project and we now refer to general methods of simulating experiments as *Monte Carlo methods*. Such methods are the focus of this chapter. The concept of simulation is quite easy to understand with simple examples.

- We could simulate the rolling of a die by using Minitab to randomly generate whole numbers between 1 and 6 inclusive, provided that the computer selects from the numbers 1, 2, 3, 4, 5, and 6 in such a way that those outcomes are equally likely.
- We could simulate births by flipping a coin, where "heads" represents a baby girl and "tails" represents a baby boy. We could also simulate births by using Minitab to randomly generate 1s (for baby girls) and 0s (for baby boys).

It is extremely important to construct a simulation so that it behaves just like the real procedure. The following example illustrates a right way and a wrong way.

EXAMPLE Describe a procedure for simulating the rolling of a pair of dice.

SOLUTION In the procedure of rolling a pair of dice, each of the two dice yields a number between 1 and 6 (inclusive) and those two numbers are then added. Any simulation should do exactly the same thing.

Right way to simulate rolling two dice: Randomly generate one number between 1 and 6, then randomly generate another number between 1 and 6, then add the two results.

Wrong way to simulate rolling two dice: Randomly generate numbers between 2 and 12. This procedure is similar to rolling dice in the sense that the results are always between 2 and 12, but these outcomes between 2 and 12 are equally likely. With real dice, the values between 2 and 12 are *not* equally likely. This simulation would yield terrible results.

4-2   Minitab Simulation Tools

Minitab includes several different tools that can be used for simulations.

1.   Click on the main menu item of **Calc**.

2.   Click on the subdirectory item of **Random Data**.

You will get a long list of different options, but we will use four of them (integer, uniform, Bernouli, normal) that are particularly relevant to the simulation techniques of Chapter4 in the textbook.

- **Integer:** Generates a sample of randomly selected integers with a specified minimum and maximum. This Minitab tool is particularly good for a variety of different simulations. Shown below is the dialog box that is provided when you select **Calc/Random Data/ Integer**. The entries in this box tell Minitab to randomly generate 500 values between 1 and 6 inclusive, and put them in column C1. This simulates rolling a single die 500 times.

- **Uniform:** Select **Calc**, then **Random Data**, then **Uniform.** The generated values are "uniform" in the sense that all possible values have the same chance of being selected. The Minitab dialog box shown below has entries indicating that 500 values should be randomly selected between 0.0 and 9.0. If you want whole numbers, use the Integer distribution instead of the Uniform distribution.

Uniform Distribution   ×

Generate [500]   rows of data

Store in column(s):

C1

Lower endpoint: [0.0]
Upper endpoint: [9.0]

Select

Help   OK   Cancel

- **Bernouli:** Select **Calc**, then **Random Data**, then **Bernouli**. The dialog box below has entries that result in a column of 0s (failure) and 1s (success), where the probability of success is 0.5. This simulates flipping a coin 500 times. The entries can be changed for different circumstances.

Bernoulli Distribution   ×

Generate [500]   rows of data

Store in column(s):

C1

Probability of success: [.5]

Select

Help   OK   Cancel

- **Normal:** Select **Calc**, then **Random Data**, then **Normal** to generate a sample of data randomly selected from a population having a normal distribution. (Normal distributions are discussed in Chapter 6 of the Triola textbook. For now, think of a normal distribution as one with a histogram that is bell–shaped.) You must enter the population mean and standard deviation. The entries in the dialog box below result in a simulated sample of 500 IQ scores taken from a population with a bell-shaped distribution that has a mean of 100 and a standard deviation of 15. Shown below the dialog box is a Minitab-generated histogram displaying the bell-shaped nature of the generated data.

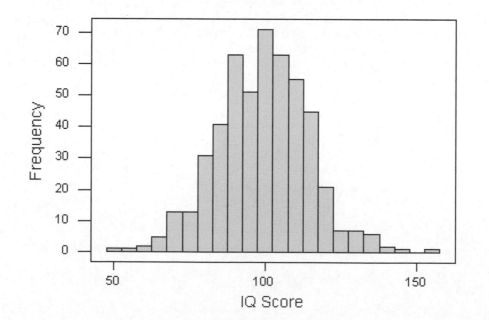

4-3 Simulation Examples

We will now illustrate the preceding Minitab features by describing specific simulations.

Simulation 1: Generating 50 births (boys/girls)

To simulate 50 births with the assumption that boys and girls are equally likely, use either of the following:

- Use Minitab's **Integer** distribution (see Section 4-2 of this manual/workbook) to generate 50 integers between 0 and 1. If you arrange the results in order (using the **Sort** feature as described in Section 3-4 of this manual/workbook), it is very easy to count the number of 0s (or boys) and the number of 1s (or girls).

- Use Minitab's **Bernouli** distribution. Enter 50 for the sample size and enter 0.5 for the probability. Again, it is very easy to count the number of 0s (or boys) and the number of 1s (or girls) if the data are sorted, as described in Section 3-4 of this manual/workbook.

Simulation 2: Rolling a *single* die 60 times

To simulate 60 rolls of a single die, use Minitab's **Integer** distribution to generate 60 integers between 1 and 6. Again, sorting the generated values makes it easy to count the number of 1s, 2s, and so on.

Simulation 3: Rolling a *pair* of dice 750 times

To simulate the rolling of a *pair* of dice 750 times, use column C1 for one die, use column C2 for the second die, then let column C3 be the *sum* of columns C1 and C2. That is, create column C1 by using the method described in Simulation 2 above. Then create column C2 by again using the method described in Simulation 2 above. Then use **Calc/Calculator** to enter the expression C1 + C2 with the result stored in column C3.

Simulation 4: Generating 25 birth dates

Instead of generating 25 results such as "January 1," or "November 27," randomly generate 25 integers between 1 and 365 inclusive. (We are ignoring leap years). Use Minitab's **Integer** distribution and enter 25 for the sample size. Also enter 1 for the minimum and enter 365 for the maximum. If you sort the simulated birth dates by using the procedure in Section 3-4 of this manual/workbook, it becomes easy to scan the sorted list and determine whether there are two birth dates that are the same. If there are two birth dates that are the same, they will show up as *consecutive* equal values in the sorted list.

CHAPTER 4 EXPERIMENTS: Probabilities Through Simulations

4—1. ***Birth Simulation*** Use Minitab to simulate 500 births, where each birth results in a boy or girl. Sort the results, count the number of girls, and enter that value here:_____

Based on that result, estimate the probability of getting a girl when a baby is born. Enter the estimated probability here: _____

The preceding estimated probability is likely to be different from 0.5. Does this suggest that the computer's random number generator is defective? Why or why not?

_____

_____

4—2. ***Dice Simulation*** Use Minitab to simulate 1000 rolls of a pair of dice. Sort the results, then find the number of times that the total was exactly 7. Enter that value here:_____

Based on that result, estimate the probability of getting a 7 when two dice are rolled. Enter the estimated probability here:_____

How does this estimated probability compare to the computed (theoretical) probability of 1/6 or 0.167? _____

_____

4—3. ***Probability of at Least 11 Girls***
    a. Use Minitab to simulate 20 births. Does the result consist of at least 11 girls? _____
    b. Repeat part (a) nine more times and record the result from part (a) along with the other nine results here: ___ ___ ___ ___ ___ ___ ___ ___ ___ ___
    c. Based on the results from part (b), what is the estimated probability of getting at least 11 girls in 20 births? _____

4-4. ***Brand Recognition*** The probability of randomly selecting san adult who recognizes the brand name of McDonald's is 0.95 (based on data from Franchise Advantage. Use Minitab to develop a simulation so that each individual outcome should be an indication of one of two results: (1) The consumer recognizes the brand name of McDonald's; (2) the consumer does not recognize the brand name of McDonald's. Conduct the simulation 1000 times and record the number of consumers who recognize the brand name of McDonald's. If possible, obtain a printed copy of the results. Is the proportion of those who recognize McDonald's reasonably close to the value of 0.95?

_____

_____

4-5.   ***Simulating Hybridization*** When Mendel conducted his famous hybridization experiments, he used peas with green pods and yellow pods. One experiment involved crossing peas in such a way that 75% of the offspring peas were expected to have green pods, and 25% of the offspring peas were expected to have yellow pods. Use Minitab to simulate 1000 peas in such a hybridization experiment. Each of the 1000 individual outcomes should be an indication of one of two results: (1) The pod is green; (2) the pod is yellow. Is the percentage of yellow peas from the simulation reasonably close to the value of 25%?

_____

_____

4—6.   ***Probability of at Least 55 Girls*** Use Minitab to conduct a simulation for estimating the probability of getting at least 55 girls in 100 births. Enter the estimated probability here:_____   Describe the procedure used to obtain the estimated probability.

_____

In testing a gender-selection method, assume that the Biogene Technology Corporation conducted an experiment with 100 couples who were treated, and that the 100 births included at least 55 girls. What should you conclude about the effectiveness of the treatment?

_____

4—7.   ***Probability of at Least 65 Girls*** Use Minitab to conduct a simulation for estimating the probability of getting at least 65 girls in 100 births. Enter the estimated probability here:_____
Describe the procedure used to obtain the estimated probability.

_____

In testing a gender-selection method, if the Biogene Technology Corporation conducted an experiment with 100 couples who were treated, and the 100 births included at least 65 girls, what should you conclude about the effectiveness of the treatment?

_____

4-8.   ***Gender-Selection Method*** As of this writing, the latest results available from the Microsort YSORT method of gender-selection consist of 127 boys in 152 births. That is, among 152 sets of parents using the YSORT method for increasing the likelihood of a boy, 127 actually had boys and the other 25 had girls. Assuming that the YSORT method has no effect and that boys and girls are equally likely, use Minitab to simulate 152 births. Is it unusual to get 127 boys in 152 births? What does the result suggest about the YSORT method?

_____

_____

4—9.   ***Simulating Three Dice*** Develop a simulation for rolling three dice. Simulate the rolling of the three dice 100 times. Describe the simulation, then use it to estimate the probability of getting a total of 10 when three dice are rolled.

_____

_____

4—10.   ***Simulating Left–Handedness*** Ten percent of us are left-handed. In a study of dexterity, people are randomly selected in groups of five. Develop a simulation for finding the probability of getting at least one left-handed person in a group of five. Simulate 100 groups of five. How does the probability compare to the correct result of 0.410, which can be found by using the probability rules in the textbook?

_____

_____

4—11.   ***Nasonex Treatment*** Nasonex is a nasal spray used to treat allergies. In clinical trials, 1671 subjects were given a placebo, and 2 of them developed upper respiratory tract infections. Another 2103 patients were treated with Nasonex and 6 of them developed upper respiratory tract infections. Assume that Nasonex has no effect on upper respiratory tract infections, so that the rate of those infections also applies to Nasonex users. Using the placebo rate of 2/1671, simulate groups of 2103 subjects given the Nasonex treatment, then determine whether a result of 6 upper respiratory tract infections could easily occur. Describe the results. What do the results suggest about Nasonex as a cause of upper respiratory tract infections?

_____

_____

4—12.   ***Birthdays*** Simulate a class of 25 birth dates by randomly generating 25 integers between 1 and 365. (We will ignore leap years.) Arrange the birth dates in ascending order, then examine the list to determine whether at least two birth dates are the same. (This is easy to do, because any two equal integers must be next to each other.)

Generated "birth dates:"   ___ ___ ___ ___ ___ ___ ___ ___ ___ ___ ___ ___ ___

___ ___ ___ ___ ___ ___ ___ ___ ___ ___ ___ ___

Are at least two of the "birth dates" the same? _____

4—13. *Birthdays* Repeat the preceding experiment nine additional times and record all ten of the yes/no responses here:

____ ____ ____ ____ ____ ____ ____ ____ ____ ____

Based on these results, what is the probability of getting at least two birth dates that are the same (when a class of 25 students is randomly selected)? _____

4—14. *Birthdays* Repeat Experiments 4-12 and 4-13 for 50 people instead of 25. Based on the results, what is the estimated probability of getting at least two birth dates that are the same (when a class of 50 students is randomly selected)? _____

4—15. *Birthdays* Repeat Experiments 4-12 and 4-13 for 100 people instead of 25. Based on the results, what is the estimated probability of getting at least two birth dates that are the same (when a class of 100 students is randomly selected)? _____

4—16. *Normally Distributed Heights* Simulate 1000 heights of adult women. (Adult women have normally distributed heights with a mean of 63.6 in. and a standard deviation of 2.5 in.) Arrange the data in ascending order, then examine the results and estimate the probability of a randomly selected woman having a height between 64.5 in. and 72 in. (Those were the height restrictions for women to fit into Russian Soyuz spacecraft when NASA and Russia ran joint missions.) Enter the estimated probability here: _____

4—17. *Normally Distributed Bulb Lives* The lifetimes of 75-watt light bulbs manufactured by the Lectrolyte Company have a mean of 1000 hours and a standard deviation of 100 hours. Generate a normally distributed sample of 500 bulb lifetimes by using the given mean and standard deviation. Examine the sorted results to estimate the probability of randomly selecting a light bulb that lasts between 850 hours and 1150 hours. Enter the result here. _____

4—18. *Normally Distributed IQ Scores* IQ scores are normally distributed with a mean of 100 and a standard deviation of 15. Generate a normally distributed sample of 800 IQ scores by using the given mean and standard deviation. Sort the results (arrange them in ascending order).

   a. Examine the sorted results to estimate the probability of randomly selecting someone with an IQ score between 90 and 110 inclusive. Enter the result here. _____
   b. Examine the sorted results to estimate the probability of randomly selecting someone with an IQ score greater than 115. _____
   c. Examine the sorted results to estimate the probability of randomly selecting someone with an IQ score less than 120. _____
   d. Repeat part a of this experiment nine more times and list all ten probabilities here.

   ____ ____ ____ ____ ____ ____ ____ ____ ____ ____

(continued)

e. Examine the ten probabilities obtained above and comment on the *consistency* of the results.

_____

f. How might we modify this experiment so that the results can become more consistent?

_____

g. If the results appear to be very consistent, what does that imply about any individual sample result?

_____

4—19. *Law of Large Numbers* In this experiment we test the Law of Large Numbers, which states that "as an experiment is repeated again and again, the empirical probability of success tends to approach the actual probability." We will use a simulation of a single die, and we will consider a success to be the outcome of 1. (Based on the classical definition of probability, we know that $P(1) = 1/6 = 0.167$.)

a. Simulate 5 trials by generating 5 integers between 1 and 6. Count the number of 6s that occurred and divide that number by 5 to get the empirical probability.
Based on 5 trials, $P(1) = $ _____.

b. Repeat part (a) for 25 trials. Based on 25 trials, $P(1) = $ _____.

c. Repeat part (a) for 50 trials. Based on 50 trials, $P(1) = $ _____.

d. Repeat part (a) for 500 trials. Based on 500 trials, $P(1) = $ _____.

e. Repeat part (a) for 1000 trials. Based on 1000 trials, $P(1) = $ _____.

f. In your own words, generalize these results in a restatement of the Law of Large Numbers.

_____

_____

_____

_____

4—20. ***Sticky Probability Problem***   Consider the following exercise, which is extremely difficult to solve with the formal rules of probability.

> *Two points along a straight stick are randomly selected. The stick is then broken at these two points. Find the probability that the three pieces can be arranged to form a triangle.*

Instead of attempting a solution using formal rules of probability, we will use a simulation. The length of the stick is irrelevant, so assume it's one unit long and its length is measured from 0 at one end to 1 at the other end. Use Minitab to randomly select the two break points with the random generation of two numbers from a uniform distribution with a minimum of 0, a maximum of 1, and 4 decimal places. Plot the break points on the "stick" below.

0_____1

A triangle can be formed if the longest segment is less than 0.5, so enter the lengths of the three pieces here: _____   _____   _____

Can a triangle be formed?

Now repeat this process nine more times and summarize all of the results below.

Trial	Break Points		Triangle formed?
1			
2			
3			
4			
5			
6			
7			
8			
9			
10			

Based on the ten trials, what is the estimated probability that a triangle can be formed? _____ This estimate gets better with more trials.

5

Probability Distributions

5-1 Exploring Probability Distributions

Chapter 5 in the Triola textbook discusses probability distributions, and its focus is *discrete* probability distributions only. These important definitions are introduced:

Definitions

A **random variable** is a variable (typically represented by x) that has a single numerical value, determined by chance, for each outcome of a procedure.

A **probability distribution** is a graph, table, or formula that gives the probability for each value of the random variable.

When working with a probability distribution, we should consider the same important characteristics introduced in Chapter 2:

1. **Center:** Measure of center, which is a representative or average value that gives us an indication of where the middle of the data set is located

2. **Variation:** A measure of the amount that the values vary among themselves

3. **Distribution:** The nature or shape of the distribution of the data, such as bell-shaped, uniform, or skewed

4. **Outliers:** Sample values that are very far away from the vast majority of the other sample values

5. **Time:** Changing characteristics of the data over time

Section 5-1 of this manual/workbook addresses these important characteristics for probability distributions. The characteristics of center and variation are addressed with formulas for finding the mean, standard deviation, and variance of a probability distribution. The characteristic of distribution is addressed through the graph of a probability histogram.

Table 5-1 Although Minitab is not designed to deal directly with a probability distribution, it can often be used. Consider Table 5-1 from the textbook, reproduced here. This table lists the probabilities for the number of peas with green pods among 5 offspring peas. This table is based on Gregor Mendel's famous hybridization experiments. If two parent peas having the green/yellow pair of genes are crossed so that 5 offspring peas are generated, there is a 0.001 probability that none of the offspring will have green pods, there is a 0.015 probability that 1 of the offspring peas will have green pods, and so on.

Table 5-1

x	$P(x)$
0	0.001
1	0.015
2	0.088
3	0.264
4	0.396
5	0.237

If you examine the data in the table, you can verify that a probability distribution is defined because the two key requirements are satisfied: (1) The sum of the probabilities is equal to 1; (2) each individual probability is between 0 and 1 inclusive. (The sum of the probabilities is actually 1.001, but it can be considered to be 1 when rounding errors are taken into account.)

Having determined that the above table does define a probability distribution, let's now see how we can use Minitab to find the mean μ and standard deviation σ. Although Minitab is not designed to directly calculate the value of the mean, there is a way to calculate exact values, and there is also a way to get very good estimated values.

Finding Exact Values of μ and σ for a Probability Distribution

1. Enter the values of the random variable x in column C1.

2. Enter the corresponding probabilities in column C2.

3. Calculate the mean μ by using Formula 5-1 in the textbook for the mean of a probability distribution: $\mu = \sum[x \cdot P(x)]$. This is accomplished by clicking on **Calc**, selecting **Calculator**, and entering the expression sum(C1*C2) with the result stored in column C3, as shown in the screen displayed below.

4. Now proceed to calculate the value of the standard deviation σ by using Formula 5-4 from the textbook: $\sigma = \sqrt{\left[\Sigma x^2 \cdot P(x)\right] - \mu^2}$. This is accomplished by selecting **Calc**, then **Calculator**, and entering the expression

$$sqrt(sum(C1**2*C2) - C3**2)$$

with the result stored in column C4.

If you use the above procedure with Table 5-1, the result will be as shown below. We can see that μ = 3.752 (or 3.8 after rounding) and the standard deviation is σ = 0.962547 (or 1.0 after rounding). The names of the columns in the display below were manually entered.

↓	C1	C2	C3	C4
	x	P(x)	Mean	St.Dev.
1	0	0.001	3.752	0.962547
2	1	0.015		
3	2	0.088		
4	3	0.264		
5	4	0.396		
6	5	0.237		

Estimating Values of μ and σ for a Probability Distribution

The above procedure for exact results is messy and somewhat difficult to remember. An easier approach for finding the mean μ and standard deviation σ is to collect a large random sample from the probability distribution, then calculate the mean and standard deviation by using **Stat**, then **Basic Statistics**, then **Display Descriptive Statistics**. Here is the Minitab procedure.

1. Enter the values of the random variable x in column C1.

2. Enter the corresponding probabilities in column C2.

3. Select **Calc**, then **Random Data**, then **Discrete**.

4. You should now get a dialog box like the one shown on the next page. The entries in this dialog box tell Minitab to generate 1,000,000 sample values randomly selected from the probability distribution defined by the values of x and $P(x)$ in columns C1 and C2. The results are stored in column C5.

5. Using the sample values in column C5, find the mean and standard deviation by using **Stat/Basic Statistics/Display Descriptive Statistics**.

Using the above procedure with Table 5-1, the author found a mean of 3.7484 (while the exact value is 3.752) and a standard deviation of 0.9701 (while the exact value is 0.952547). The simulated sample provided results that are very close to the exact values.

5-2 Binomial Distributions

A **binomial distribution** is defined in Section 5-3 of the Triola textbook to be to be a probability distribution that meets all of the following requirements:

1. The experiment must have a fixed number of trials.
2. The trials must be independent. (The outcome of any individual trial doesn't affect the probabilities in the other trials.)
3. Each trial must have all outcomes classified into two categories.
4. The probabilities must remain constant for each trial.

We also introduced notation with S and F denoting success and failure for the two possible categories of all outcomes. Also, p and q denote the probabilities of S and F, respectively, so that $P(S) = p$ and $P(F) = q$. We also use the following symbols.

n denotes the fixed number of trials

x denotes a specific number of successes in n trials so that x can be any whole number between 0 and n, inclusive.

p denotes the probability of success in *one* of the n trials.

q denotes the probability of failure in *one* of the n trials.

$P(x)$ denotes the probability of getting exactly x successes among the n trials.

Section 5-3 in the Triola textbook describes three methods for determining probabilities in binomial experiments. Method 1 uses the binomial probability formula:

$$P(x) = \frac{n!}{(n - x)!x!} p^x q^{n-x}$$

Method 2 requires computer usage. We noted in the textbook that if a computer and software are available, this method of finding binomial probabilities is fast and easy, as shown in the following Minitab procedure.

Minitab Procedure for Finding Probabilities with a Binomial Distribution

1. In column C1, enter the values of the random variable x for which you want probabilities. (You can enter all values of x as 0, 1, 2, . . .)

2. Click on **Calc** from the main menu.

3. Select **Probability Distributions**.

4. Select **Binomial**.

5. The entries in this dialog box shown below correspond to a binomial distribution with 5 trials, and a probability of success equal to 0.75. The probability values are to be computed for the values of x that have been entered in column C1, and the results will be stored in column C2.

Binomial Distribution dialog box:

C1　x
C2　P(x)

- Probability
- Cumulative probability
- Inverse cumulative probability

Number of trials:　5
Event probability:　0.75

- Input column:　c1
  Optional storage:　c2
- Input constant:
  Optional storage:

Select　　Help　　OK　　Cancel

Consider the following example.

> ***Pea Hybridization***   Use the binomial probability formula to find the
> probability of getting exactly 4 peas with green pods when 5 offspring peas
> are generated under the assumption that the probability of a pea with a
> green pod is 0.75. That is, find the value $P(4)$ given that $n = 5$, $x = 4$, $p =$
> 0.75, and $q = 0.25$.

Begin by entering the values of 0, 1, 2, 3, 4, and 5 in column C1, and continue with the above
procedure. Columns C1 and C2 will display the values shown below. This display corresponds to
the values in Table 5-1, with the probabilities rounded to 6 decimal places instead of 3 (as in
Table 5-1). We can easily see from this display that $P(4) = 0.395508$, which is the same value
obtained by using the binomial probability formula. (*Caution*: Don't use the Minitab row
numbers in the column at the extreme left for the values of x; be sure to use the values in column
C1 for the values of x.) The table of binomial probabilities (Table A-1) in Appendix B of the
textbook includes limited values of n and p, but Minitab is much more flexible in the values of n
and p that can be used. Users of Minitab can therefore ignore Table A-1 and all of its inherent
limitations.

x	P(x)
0	0.000977
1	0.014648
2	0.087891
3	0.263672
4	0.395508
5	0.237305

5-3   Poisson Distributions

Textbooks in the Triola Statistics Series (except for *Essentials of Statistics*) discuss the Poisson
distribution. A Poisson distribution is a discrete probability distribution that applies to
occurrences of some event *over a specified interval*. The random variable x is the number of
occurrences of the event in an interval, such as time, distance, area, volume, or some similar unit.
The probability of the event occurring x times over an interval is given by this formula:

$$P(x) = \frac{\mu^x \cdot e^{-\mu}}{x!} \qquad \text{where } e \approx 2.71828$$

The textbook also notes that the Poisson distribution is sometimes used to approximate
the binomial distribution when $n \geq 100$ and $np \leq 10$; in such cases, we use $\mu = np$. If using
Minitab, the Poisson approximation to the binomial distribution isn't used much, because we can
easily find binomial probabilities for a wide range of values for n and p, so an approximation is
not necessary.

Minitab Procedure for Finding Probabilities for a Poisson Distribution

1. Determine the value of the mean μ.

2. In column C1, enter the values of the random variable x for which you want probabilities. (You can enter values of x as 0, 1, 2, ...)

3. Click on **Calc** from the main menu.

4. Select **Probability Distributions**.

5. Select **Poisson**.

6. You will now see a dialog box like the one shown below. The probability values are to be computed for the values of x that have been entered in column C1, and the results will be stored in column C2.

Consider the following example based on Example 1 from the textbook:

> *Earthquakes* For a recent period of 100 years, there were 93 major earthquakes (measuring at least 6.0 on the Richter scale) in the world (based on data from the *World Almanac and Book of Facts*). Assume that the Poisson distribution is a suitable model. If $P(x)$ is the probability of x earthquakes in a randomly selected year, find $P(0)$, $P(1)$, $P(2)$, $P(3)$, $P(4)$, $P(5)$, $P(6)$, and $P(7)$.

The mean number of earthquakes per year is

$$\mu = \frac{\text{number of earthquakes}}{\text{number of years}} = \frac{93}{100} = 0.93$$

Having found the required mean μ, we can now proceed to use Minitab as described above. If you enter the x values of 0, 1, 2, 3, 4, 5, 6, and 7 in column C1, you will get results in columns C1 and C2 as shown below. We can see that the probability of 0 earthquakes in a year is 0.394554, the probability of 1 earthquake in a year is 0.366935, and so on.

C1	C2
x	P(x)
0	0.394554
1	0.366935
2	0.170625
3	0.052894
4	0.012298
5	0.002287
6	0.000355
7	0.000047

5-4 Cumulative Probabilities

The main objective of this section is to stress the point that *cumulative probabilities* are often critically important. By *cumulative* probability, we mean the probability that the random variable *x* has a range of values instead of a single value. Here are typical examples:

- Find the probability of getting *at least* 13 girls in 14 births.
- Find the probability of *more than* 5 wins when roulette is played 200 times.

A *cumulative* probability is much more important than the probability of any individual event. Section 5-2 of the textbook includes criteria for determining when results are unusual, and the following criteria involve cumulative probabilities.

Using Probabilities to Determine When Results Are Unusual

- **Unusually high number of successes:** *x* successes among *n* trials is an *unusually high* number of successes if $P(x \text{ or more}) \leq 0.05^{*}$.

- **Unusually low number of successes:** *x* successes among *n* trials is an *unusually low* number of successes if $P(x \text{ or fewer}) \leq 0.05^{*}$.

*The value of 0.05 is commonly used, but is not absolutely rigid. Other values, such as 0.01, could be used to distinguish between events that can easily occur by chance and events that are very unlikely to occur by chance.

Example 7 in Section 5-2 of the textbook includes this: Use *probabilities* to determine whether 1 is an unusually low number of peas with green pods when 5 offspring are generated from parents both having the green/yellow pair of genes. The probabilities are listed in Table 5-1. In addressing this issue, we find that the probability of 1 or fewer peas with green pods is 0.016, which is small (such as less than 0.05), so we conclude that it *is* unusual to get only 1 pea with green pods. Also, the textbook stresses the important role of cumulative probabilities with an example, reproduced here because it is so important:

Suppose you were tossing a coin to determine whether it favors heads, and suppose 1000 tosses resulted in 501 heads. This is not evidence that the coin favors heads, because it is very easy to get a result like 501 heads in 1000 tosses just by chance. Yet, the probability of getting *exactly* 501 heads in 1000 tosses is actually quite small: 0.0252. This low probability reflects the fact that with 1000 tosses, *any specific* number of heads will have a very low probability. However, we do not consider 501 heads among 1000 tosses to be *unusual*, because the probability of *501 or more* heads is high: 0.487.

Cumulative probabilities therefore play a critical role in identifying results that are considered to be *unusual*. Later chapters focus on this important concept. If you examine the dialog boxes used for finding binomial probabilities and Poisson probabilities, you can see that in addition to obtaining probabilities for specific numbers of successes, you can also obtain *cumulative* probabilities.

CHAPTER 5 EXPERIMENTS: Probability Distributions

5-1. A probability distribution is described below. Use Minitab with the procedure described in Section 5-1 of this manual/workbook for the following.

  a. Using the *exact* procedure, find the mean and standard deviation.
   Mean:_____ St. Dev.:_____

  b. Using the *approximate* procedure, estimate the mean and standard deviation.
   Mean:_____ St. Dev.:_____

  c. Compare the exact values to the estimated values.

Genetic Disorder Three males with an X–linked genetic disorder have one child each. The random variable x is the number of children among the three who inherit the X–linked genetic disorder, and the probability distribution for that number of children is given in the table below.

x	$P(x)$
0	0.4219
1	0.4219
2	0.1406
3	0.0156

5-2. A probability distribution is described below. Use Minitab with the procedure described in Section 5-1 of this manual/workbook for the following.

  a. Using the *exact* procedure, find the mean and standard deviation.
   Mean:_____ St. Dev.:_____

  b. Using the *approximate* procedure, estimate the mean and standard deviation.
   Mean:_____ St. Dev.:_____

  c. Compare the exact values to the estimated values.

Air America has a policy of routinely overbooking flights. The random variable x represents the number of passengers who cannot be boarded because there are more passengers than seats (based on data from an IBM research paper by Lawrence, Hong, and Cherrier).

x	$P(x)$
0	0.051
1	0.141
2	0.274
3	0.331
4	0.187

5-3.   A probability distribution is described below. Use Minitab with the procedure described in Section 5-1 of this manual/workbook for the following.

    a.   Using the *exact* procedure, find the mean and standard deviation.
        Mean:_____ St. Dev.:_____

    b.   Using the *approximate* procedure, estimate the mean and standard deviation.
        Mean:_____ St. Dev.:_____

    c.   Compare the exact values to the estimated values.

_____

_____

_____

Groups of five babies are randomly selected. In each group, the random variable x is the number of babies with green eyes (base on data from a study by Dr. Sorita Soni at Indiana University). (The symbol 0+ denotes a positive probability value that is very small.)

x	$P(x)$
0	0.528
1	0.360
2	0.098
3	0.013
4	0.001
5	0+

5-4.   ***Binomial Probabilities***   Use Minitab to find the binomial probabilities corresponding to $n = 4$ and $p = 0.05$. Enter the results below, along with the corresponding results found in Table A-1 of the textbook.

x	$P(x)$ from Minitab	$P(x)$ from Table A-1
0		
1		
2		
3		
4		

By comparing the above results, what advantage does Minitab have over Table A-1?

_____

_____

5-5.   ***Binomial Probabilities***   Consider the binomial probabilities corresponding to $n = 4$ and $p = 1/4$ (or 0.25). If you attempt to use Table A-1 for finding the probabilities, you will find that the table does not apply to a probability of $p = 1/4$. Use Minitab to find the probabilities and enter the results below.

$P(0) = $ _____

$P(1) = $ _____

$P(2) = $ _____

$P(3) = $ _____

$P(4) = $ _____

5-6.   ***Binomial Probabilities***   Assume that boys and girls are equally likely and 100 births are randomly selected. Use Minitab with $n = 100$ and $p = 0.5$ to find $P(x)$, where x represents the number of girls among the 100 babies.

$P(35) = $ _____

$P(45) = $ _____

$P(50) = $ _____

5-7.   ***Binomial Probabilities***   We often assume that boys and girls are equally likely, but the actual values are $P(\text{boy}) = 0.5121$ and $P(\text{girl}) = 0.4879$. Repeat Experiment 5-6 using these values, then compare these results to those obtained in Experiment 5-6.

How does the result compare to the result from Experiment 5-6?

$P(35) = $ _____   _____

_____

$P(45) = $ _____   _____

_____

$P(50) = $ _____   _____

_____

5-8.   ***Cumulative Probabilities***   Assume that P(boy) = 0.5121, P(girl) = 0.4879, and that 100 births are randomly selected. Use Minitab to find the probability that the number of girls among 100 babies is . . .

    a.   Fewer than 60   _____

    b.   Fewer than 48   _____

    c.   At most 30   _____

    d.   At least 55   _____

    e.   More than 40   _____

5-9.   ***Identifying 0+***   In Table A-1 from the textbook, the probability corresponding to $n = 12$, $p = 0.10$, and $x = 6$ is shown as 0+. Use Minitab to find the corresponding probability and enter the result here.   _____

5-10.   ***Identifying 0+***   In Table A-1 from the textbook, the probability corresponding to $n = 15$, $p = 0.80$, and $x = 5$ is shown as 0+. Use Minitab to find the corresponding probability and enter the result here.   _____

5-11.   ***Identifying a Probability Distribution***   Use Minitab to construct a table of x and $P(x)$ values corresponding to a binomial distribution in which $n = 26$ and $p = 0.3$. Enter the table in the space below.

    Table:

Exercises 5–12 through 5-16 involve binomial distributions. Use Minitab for those exercises.

5-12. ***Brand Recognition*** The brand name of Mrs. Fields (cookies) has a 90% recognition rate (based on data from Franchise Advantage). If Mrs. Fields herself wants to verify that rate by beginning with a small sample of ten randomly selected consumers, find the probability that exactly nine of the ten consumers recognize her brand name. Also find the probability that the number who recognize her brand name is *not* nine.

5-13. ***Brand Recognition*** The brand name of McDonald's has a 95% recognition rate (based on data from Retail Marketing Group). If a McDonald's executive wants to verify that rate by beginning with a small sample of 15 randomly selected consumers, find the probability that exactly 13 of the 15 consumers recognize the McDonald's brand name. Also find the probability that the number who recognize the brand name is *not* 13.

5-14. ***Eye Color*** In the United States, 40% of the population have brown eyes (based on data from Dr. P Sorita Soni at Indiana University). If 14 people are randomly selected, find the probability that at least 12 of them have brown eyes. Is it unusual to randomly select 14 people and find that at least 12 of them have brown eyes? Why or why not?

5-15. ***Credit Rating*** There is a 1% delinquency rate for consumers with FICO (Fair Isaac & Company) credit rating scores above 800. If the Jefferson Valley Bank provides large loans to 12 people with FICO scores above 800, what is the probability that at least one of them becomes delinquent? Based on that probability, should the bank plan on dealing with a delinquency?

5-16. ***Genetics*** Ten peas are generated from parents having the green/yellow pair of genes, so there is a 0.75 probability that an individual pea will have a green pod. Find the probability that among the 10 offspring peas, at least nine have green pods. Is it unusual to get at least nine peas with green pods when ten offspring peas are generated? Why or why not?

Exercises 5-17 through 5-20 involve Poisson distributions. Use Minitab for those exercises.

5–17. ***Radioactive Decay*** Radioactive atoms are unstable because they have too much energy. When they release their extra energy, they are said to decay. When studying Cesium 137, it is found that during the course of decay over 365 days, 1,000,000 radioactive atoms are reduced to 977,287 radioactive atoms.
    a.   Find the mean number of radioactive atoms lost through decay in a day. _____
    b.   Find the probability that on a given day, 50 radioactive atoms decayed. _____

5-18. ***Disease Cluster*** Neuroblastoma, a rare form of malignant tumor, occurs in 11 children in a million, so its probability is 0.000011. Four cases of neuroblastoma occurred in Oak Park, Illinois, which had 12,429 children.
    a.   Assuming that neoroblastoma occurs as usual, find the mean number of cases in groups of 12,429 children. _____
    b.   Find the probability that the number of neuroblastoma cases in a group of 12,429 children is 0 or 1. _____
    c.   What is the probability of more than one case of neuroblastoma? _____
    d.   Does the cluster of four cases appear to be attributable to random chance? Why or why not?

5-19. ***Deaths From Horse Kicks*** A classic example of the Poisson distribution involves the number of deaths caused by horse kicks of men in the Prussian Army between 1875 and 1894. Data for 14 corps were combined for the 20-year period, and the 280 corps-years included a total of 196 deaths. After finding the mean number of deaths per corps-year, find the probability that a randomly selected corps-year has the following numbers of deaths.
    a. 0 _____   b. 1 _____   c. 2 _____   d. 3 _____   e. 4 _____
The actual results consisted of these frequencies: 0 deaths (in 144 corps-years); 1 death (in 91 corps-years); 2 deaths (in 32 corps-years); 3 deaths (in 11 corps-years); 4 deaths (in 2 corps-years). Compare the actual results to those expected from the Poisson probabilities. Does the Poisson distribution serve as a good device for predicting the actual results?

5-20. ***Life Insurance*** A Fidelity life insurance company charges $226 for a $50,000 life insurance policy for a 50-year-old female. The probability that such a female survives the year is 0.9968 (based on data from the U. S. Department of Health and Human Services). Assume that the company sells 700 such policies to 50-year-old females, so it collects $158,200 in policy payments. The company will make a profit if fewer than four of the 700 women die during the year.
    a.   What is the mean number of deaths in such groups of 700 females? _____
    b.   Find the probability that the company makes a profit from the 700 policies. Is that probability high enough so that the company is almost sure to make a profit?

6

Normal Distributions

6-1 Finding Areas and Values with a Normal Distribution

The Triola textbook describes methods for working with standard and nonstandard normal distributions. (A **standard normal distribution** has a mean of 0 and a standard deviation of 1.) Table A-2 in Appendix B of the textbook lists a wide variety of different *z* scores along with their corresponding areas. Minitab can also be used to find probabilities associated with the normal distribution. Although Table A-2 in the textbook includes only limited values of *z* scores and their corresponding areas, Minitab can be used with infinitely many different possible choices. In the following procedure, note that like Table A-2 in the textbook, Minitab areas are to the *left* of the related *x* value. Here is the procedure for finding areas similar to those in the body of Table A-2 from the textbook.

Minitab Procedure for Finding Areas and Values with a Normal Distribution

1. Select **Calc** from the main menu at the top of the screen.

2. Select **Probability Distributions** from the subdirectory.

3. Select **Normal**.

4. You will get a dialog box like the one shown below.

When making choices and entries in the dialog box, consider the following:

- **Probability density** is the height of the normal distribution curve, so we will rarely use this option.

- **Cumulative probability:** Use this option to find an area or probability. Note that you will get the area under the curve to the *left* of the x value(s) entered in column C1. (You must have already entered the x value(s) in column C1, or you could enter a constant value of x by selecting the **Input constant** option.)

- **Inverse cumulative probability:** Use this option to find a value corresponding to a given area. The resulting x value(s) separates an area to the *left* that is specified in column C1. (You must have already entered the cumulative left areas in column C1, or you could enter a single area as a constant by selecting the **Input constant** option.)

The input column is the column containing values (either areas or x values) that you have already entered, and the output column is the column that will list the results (either x values or areas).

5. Click **OK** and the results will be listed in the output column.

Examples: Assume that IQ scores are normally distributed with a mean of 100 and a standard deviation of 15.

- **Finding Area:** To find the area to the left of 115, enter 115 in column C1, then select **Calc, Probability Distributions**, then **Normal**. In the dialog box, select the option of **Cumulative Probability** (because you want an *area*). The result will be 0.843415, which is the area to the *left* of 115.

- **Finding a Value:** To find the 90th percentile, enter 0.9 in column C1, then select **Calc, Probability Distributions**, then **Normal**. In the dialog box, select the option of **Inverse Cumulative Probability** (because you want a *value* of IQ score). The result will be 119.223, which is the IQ score separating an area of 0.9 to its *left*.

6-2 Simulating and Generating Normal Data

We can learn much about the behavior of normal distributions by analyzing samples obtained from them. Sampling from real populations is often time consuming and expensive, but we can use the wonderful power of computers to obtain samples from theoretical normal distributions, and Minitab has such a capability, as described below. Let's consider IQ scores. IQ tests are designed to produce a mean of 100 and a standard deviation of 15, and we expect that such scores are normally distributed. Suppose we want to learn about the variation of sample means for samples of IQ scores. Instead of going out into the world and randomly selecting groups of people and administering IQ tests, we can sample from theoretical populations. We can then learn much about the distribution of sample means. The following procedure allows you to obtain a random sample from a normally distributed population with a given mean and standard deviation.

Generating a Random Sample from a Normally Distributed Population

1. Click on the main menu item of **Calc**.
2. Click on **Random Data**.
3. Click on **Normal**.
4. You will now get a dialog box such as the one shown below. You must enter the population mean and standard deviation. The entries in the dialog box below result in simulated sample of 500 IQ scores taken from a population with a normal distribution that has a mean of 100 and a standard deviation of 15. The sample values are stored in column C1.

Important: *When using the above procedure, the generated sample will not have a mean equal to the specified population mean, and the standard deviation of the sample will not have a value equal to the specified standard deviation.* When the author generated 500 sample values with the above dialog box, he obtained sample values with a mean of 100.28 and a standard deviation of 13.98. Remember, you are sampling from a population with the specified parameters; you are not generating a sample with statistics exactly equal to the specified values. See Section 6–5 of this manual/workbook for a procedure that can be used to generate a sample with exact values of a specified mean and standard deviation.

6-3 The Central Limit Theorem

The textbook introduces the central limit theorem, which is then used in subsequent chapters when the important topics of estimating parameters and hypothesis testing are discussed. Here is the statement of the central limit theorem in the Triola textbook:

The Central Limit Theorem and the Sampling Distribution of $\bar{x}$

Given:
1. The random variable x has a distribution (which may or may not be normal) with mean μ and standard deviation σ.
2. Simple random samples all of the same size n are selected from the population. (The samples are selected so that all possible samples of size n have the same chance of being selected.)

Conclusions:
1. The distribution of sample means $\bar{x}$ will, as the sample size increases, approach a *normal* distribution.
2. The mean of all sample means is the population mean μ.
3. The standard deviation of all sample means is $\sigma/\sqrt{n}$.

Practical Rules Commonly Used
1. If the original population is *not normally distributed*, here is a common guideline: For $n > 30$, the distribution of the sample means can be approximated reasonably well by a normal distribution. (There are exceptions, such as populations with very nonnormal distributions requiring sample sizes larger than 30, but such exceptions are relatively rare.) The distribution of sample means gets closer to a normal distribution as the sample size n becomes larger.
2. If the original population is *normally distributed*, then for *any* sample size n, the sample means will be normally distributed.

   Example In the study of methods of statistical analysis, it is extremely helpful to have a clear understanding of the statement of the central limit theorem. It is helpful to use Minitab in conducting an experiment that illustrates the central limit theorem. For example, let's use Minitab to generate five columns of data randomly selected with Minitab's **Integer** distribution, which generates integers so that they are all equally likely. After generating 5000 values (each between 0 and 9) in columns C1, C2, C3, C4, and C5, we will stack the 25000 values in column C6. We will also calculate the 5000 sample means of the values in columns C1, C2, C3, C4, and C5 and enter the results in column C7. (The sample means can be obtained by selecting **Calc**, then selecting **Row Statistics**; select **Mean** and enter C1-C7 for Input variables.) See the histograms shown below. The first histogram shows the distribution of the 25,000 generated integers, but the second histogram shows the distribution of the 5000 sample means. We can clearly see that the sample means have a distribution that is approximately normal, even thought the original population has a uniform distribution.

Distribution of 25,000 randomly selected integers

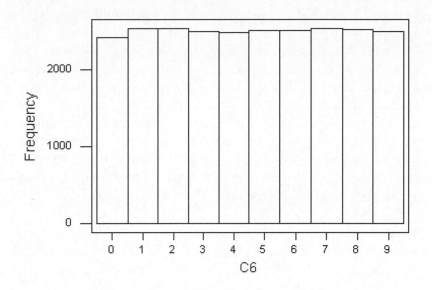

Distribution of 5000 sample means

6-4 Assessing Normality

Textbooks in the Triola statistics series discuss criteria for determining whether sample data appear to come from a population having a normal distribution. (This section of Assessing Normality is not included in *Essentials of Statistics*.) These criteria are listed:

1. **Histogram:** Construct a Histogram. Reject normality if the histogram departs dramatically from a bell shape. Minitab can generate a histogram. Section 2-2 of this manual/workbook gives the Minitab procedure for generating a histogram.

2. **Outliers:** Identify outliers. Reject normality if there is more than one outlier present. (Just one outlier could be an error or the result of chance variation, but be careful, because even a single outlier can have a dramatic effect on results.) Using Minitab, we can sort the data and easily identify any values that are far away from the majority of all other values. Minitab-generated boxplots can also be used to identify potential outliers. Section 3-3 of this manual/workbook provides the Minitab procedure for generating a "modified boxplot." Modified boxplots are described in Part 2 of Section 3-4 in the textbook. In a Minitab-generated boxplot, points identified with asterisks are potential outliers.

3. **Normal Quantile Plot:** If the histogram is basically symmetric and there is at most one outlier, construct a *normal quantile plot*. Examine the normal quantile plot and reject normality if the points do not lie close to a straight line, or if the points exhibit some systematic pattern that is not a straight-line pattern. Minitab can generate a *normal probability plot*, which can be interpreted the same way as the normal quantile plot described in the textbook. (Select **Stat**, then **Basic Statistics**, then **Normality Test**.) Also, a normal probability plot can be generated with "percentile" curves that serve as boundaries for departures from normal distributions. (Select **Graph**, then **Probability Plot**, and select the option of **Single**.)

Also, Part 2 of Section 6-7 in the Triola textbook includes a brief reference to the Ryan-Joiner test as one of several formal tests of normality, each having their own advantages and disadvantages. Minitab includes the Ryan-Joiner test, the Anderson-Darling test, and the Kolmogorov-Smirnov test as formal tests that can be used for assessing the normality of a data set. Any one of these three normality tests can be conducted by selecting **Stat**, then **Basic Statistics**, then **Normality Test**.

Normal Probability Plot To obtain a normal probability plot, click on the main menu item of **Stat**, then select **Basic Statistics**, then **Normality Test**. You will get a dialog box like the one shown below. The entries in the window at the left are present because the worksheet CHMOVIE (child movies) was opened so that columns with numerical data are listed. (See Example 4 in Section 6-7 of the textbook.) Click on the desired column so that its label appears in the "Variable" box. Click **OK**. The resulting normal quantile plot will be as shown on the next page.

Using the normal probability plot shown below, we see that the plot does *not* yield a pattern of points that reasonably approximates a straight line, so we conclude that the times for tobacco use are not normally distributed.

Shown below is the normal probability plot obtained by selecting **Graph**, **Probability Plot**, and **Single**. See that boundary lines are included. Because the points extend beyond those boundary lines, we conclude that the distribution of data is not a normal distribution.

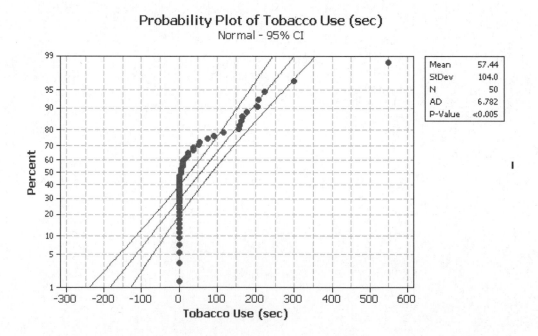

6-5 Normal Approximation to Binomial

Section 6-6 in the Triola textbook discusses the topic of using a normal distribution as an approximation to a binomial distribution. That section includes this comment:

> *Note:* Instead of using a normal distribution as an approximation to a binomial probability distribution, most practical applications of the binomial distribution can be handled with computer software or a calculator.

When working with applications involving a binomial distribution, Minitab can be used to find *exact* results, so there is no need to approximate the binomial distribution with a normal distribution. Consider Example 1 in Section 6-6 of the textbook:

> The author was mailed a survey from Viking River Cruises, and the survey included a request for an e-mail address. Assume that the survey was sent to 40,000 people and that for such surveys, the percentage of responses with an e-mail address is 3%. If the true goal of the survey was to acquire a bank of at least 1150 e-mail addresses, find the probability of getting at least 1150 responses with e-mail addresses.

Based on the above statements from Example 1 in the textbook, we have $n = 40,000$ and $p = 0.03$, and we want to find $P(x \geq 1150)$, which is the probability of getting at least 1150 responses with e-mail addresses. Using Minitab with the binomial distribution procedure described in Section 5-2 of this manual/workbook, we can find that the probability of 1149 or fewer responses with e-mail addresses is 0.0686665, so the *exact* probability of at least 1150 such responses is $1 - 0.0686665 = 0.9313335$. This is a more accurate result than the result of 0.9306 found by using the normal distribution as an approximation to the binomial distribution. For Minitab users, the method of approximating a binomial distribution with a normal distribution is generally obsolete.

CHAPTER 6 EXPERIMENTS: Normal Distributions

6-1.   *Finding Probabilities for a Normal Distribution*   Use Minitab's **Normal** probability distribution module to find the indicated probabilities. First select **Calc** from the main menu, then select **Probability Distributions**, then **Normal**. Probabilities can be found by using the **Cumulative Probability** option. Remember that like Table A-2 in the textbook, Minitab's probabilities correspond to cumulative areas from the left.

   a.   Given a population with a normal distribution, a mean of 0, and a standard deviation of 1, find the probability of a value less than 0.75._____

   b.   Given a population with a normal distribution, a mean of 100, and a standard deviation of 15, find the probability of a value less than 83._____

   c.   Given a population with a normal distribution, a mean of 120, and a standard deviation of 10, find the probability of a value greater than 135._____

   d.   Given a population with a normal distribution, a mean of 98.6, and a standard deviation of 0.62, find the probability of a value between 97.0 and 99.0._____

   e.   Given a population with a normal distribution, a mean of 150, and a standard deviation of 12, find the probability of a value between 147 and 160._____

6-2.   *Finding Values for a Normal Distribution*   Use Minitab's **Normal** module to find the indicated vales. First select **Calc** from the main menu, then select **Probability Distributions**, then **Normal**. Values can be found by using the **Inverse cumulative probability** option. Remember that like Table A-2 in the textbook, Minitab's probabilities correspond to cumulative areas from the left.

   a.   Given a population with a normal distribution, a mean of 0, and a standard deviation of 1, what value has an area of 0.4 to its left?_____

   b.   Given a population with a normal distribution, a mean of 100, and a standard deviation of 15, what value has an area of 0.3 to its right?_____

   c.   Given a population with a normal distribution, a mean of 85, and a standard deviation of 8, what value has an area of 0.95 to its left?_____

   d.   Given a population with a normal distribution, a mean of 45.5, and a standard deviation of 3.5, what value has an area of 0.88 to its right?_____

   e.   Given a population with a normal distribution, a mean of 94, and a standard deviation of 6, what value has an area of 0.01 to its left?_____

6-3.   ***Central Limit Theorem*** In this experiment we will illustrate the central limit theorem. See the example in Section 6-3 of this manual/workbook.

    a.   Use Minitab to create columns C1, C2, C3, C4, and C5 so that each column contains 250 values that are randomly generated integers between 0 and 9.

    b.   Use **Data/Stack** to stack all of the 1250 values in column C6. Obtain a printed copy of the histogram for the sample data in column C6. Also find the mean and standard deviation and enter those values here.

           Mean:_____   Standard deviation:_____

    c.   Now create a column C7 consisting of the 250 sample means. The first entry of column C7 is the mean of the first entries in columns C1 through C5, the second entry of C7 is the mean of the second entries in columns C1 through C5, and so on. (Click on **Calc**, select **Row Statistics**, select **Mean**, and enter C1-C5 for the input variables. Enter C7 as the column for storing the results. Obtain a printed copy of the histogram for the sample data in column C7. Also find the mean and standard deviation and enter those values here.

           Mean:_____   Standard deviation:_____

    d.   How do these results illustrate the central limit theorem?

           _____

           _____

6-4.   ***Central Limit Theorem*** In this experiment we will illustrate the central limit theorem. See the example in Section 6-3 of this manual/workbook.

    a.   Use Minitab to create columns C1 through C20 in such a way that each column contains 5000 values that are randomly generated from Minitab's **Uniform** distribution with a minimum of 0 and a maximum of 5.

    b.   Stack all of the sample values in column C21. [See part (b) of the preceding exercise.] Obtain a printed copy of the histogram for the sample data in column C21. Also find the mean and standard deviation and enter those values here.

           Mean:_____   Standard deviation:_____

    c.   Now create a column C22 consisting of the 5000 sample means. The first entry of column C22 is the mean of the first entries in columns C1 through C20, the second entry of C22 is the mean of the second entries in columns C1 through C20, and so on. [See part (c) of the preceding exercise.]
Obtain a printed copy of the histogram for the sample data in column C22. Also find the mean and standard deviation and enter those values below.
Mean:_____   Standard deviation:_____

d.   How do these results illustrate the central limit theorem?

_____

_____

6-5.   *Identifying Significance* People generally believe that the mean body temperature is 98.6°F. The body temperature data set in Appendix B of the textbook includes a sample of 106 body temperatures with these properties: The distribution is approximately normal, the sample mean is 98.20°F, and the standard deviation is 0.62°F. We want to determine whether these sample results differ from 98.6°F by a *significant* amount. One way to make that determination is to study the behavior of samples drawn from a population with a mean of 98.6 (and a standard deviation of 0.62°F and a normal distribution).

a.   Use Minitab to generate 106 values from a normally distributed population with a mean of 98.6 and a standard deviation of 0.62. Use **Stat/Basic Statistics/ Display Descriptive Statistics** to find the mean of the generated sample. Record that mean here:_____

b.   Repeat part a nine more times and record the 10 sample means here:

_____

c.   By examining the 10 sample means in part b, we can get a sense for how much sample means vary for a normally distributed population with a mean of 98.6 and a standard deviation of 0.62. After examining those 10 sample means, what do you conclude about the likelihood of getting a sample mean of 98.20? Is 98.20 a sample mean that could easily occur by chance, or is it significantly different from the likely sample means that we expect from a population with a mean of 98.6?

_____

_____

d.   Given that researchers did obtain a sample of 106 temperatures with a mean of 98.20°F, what does their result suggest about the common belief that the population mean is 98.6°F?

_____

_____

6-6.   ***Identifying Significance***   This experiment involves one of the data sets in Appendix B of the textbook: "Weights and Volumes of Cola." The Minitab worksheet is COLA.

    a.   Open the worksheet COLA and find the mean and standard deviation of the sample consisting of the volumes of cola in cans of regular Coke. Enter the results here.   Sample mean:_____   Standard deviation:_____

    b.   Generate 10 different samples, where each sample has 36 values randomly selected from a normally distributed population with a mean of 12 oz and a standard deviation of 0.115 oz (based on the claimed volume printed on the cans and the data in Appendix B). For each sample, record the sample mean and enter it here.

_____

    c.   By examining the 10 sample means in part b, we can get a sense for how much sample means vary for a normally distributed population with a mean of 12 and a standard deviation of 0.115. After examining those 10 sample means, what do you conclude about the likelihood of getting a sample mean like the one found for the sample volumes in Appendix B? Is the mean for the sample a value that could easily occur by chance, or is it significantly different from the likely sample means that we expect from a population with a mean of 12?

_____

    d.   Consider the sample mean found from the volumes of regular Coke listed in Appendix B from the textbook. Does it suggest that the population mean of 12 oz (as printed on the label) is not correct?

_____

*Assessing **Normality**. In the following exercises, refer to the indicated data set and use Minitab to determine whether the data have a normal distribution. Give a reason explaining your choice.*

6-7.   ***Space Shuttle Flights***   The lengths (in hours) of flights of NASA's Space Transport System (Shuttle) as listed in Data Set 10 in Appendix B. The Minitab worksheet is NASA.

6-8.   ***Astronaut Flights***   The numbers of flights by NASA astronauts, as listed in Data Set 10 in Appendix B. The Minitab worksheet is NASA.

6-9.   ***Heating Degree Days*** The values of heating degree days, as listed in Data Set 12 in Appendix B. The Minitab worksheet is ELECTRIC.

6-10.   ***Generator Voltage*** The measured voltage levels from a generator, as listed in Data Set 13 in Appendix B. The Minitab worksheet is VOLTAGE.

7

Confidence Intervals and Sample Sizes

7-1 Confidence Intervals for Estimating p

Section 7-2 of the Triola textbook presents methods for using a sample proportion to estimate a population proportion, and a confidence interval is a common and important tool used for that purpose. A **confidence interval** (or **interval estimate**) is a range (or an interval) of values used to estimate the true value of a population parameter.

Important note about methods: Section 7-2 in the textbook gives a method for constructing a confidence interval estimate of a population proportion p. The textbook procedure is based on using a normal distribution as an approximation to a binomial distribution. However, *Minitab uses an* exact *calculation instead of a normal approximation.* Minitab provides an option for using the normal approximation method, but it is better to use the exact procedure.

Finding x: Minitab can use the sample size n and the number of successes x to construct a confidence interval estimate of a population proportion. In some cases, the values of x and n are both known, but in other cases the given information may consist of n and a sample percentage. For example, suppose we know that among 1501 people surveyed, 70% said that they believe in global warming. Based on that information, we know that $n = 1501$ and $\hat{p} = 0.70$. Because $\hat{p} = x/n$, it follows that $x = \hat{p}n$, so the number of successes can be found by multiplying the sample proportion p and the sample size n. Given that $n = 1501$ and $\hat{p} = 0.70$, we find $x = (0.70)(1501) = 1050.7$, which we round to the whole number 1051. It's actually quite simple: 70% of 1501 is $0.70 \times 1501 = 1051$ (rounded to a whole number).

> **To find the number of successes x from the sample proportion and sample size:**
> **Calculate $x = \hat{p} \cdot n$ and round the result to the nearest whole number.**

After having determined the value of the sample size n and the number of successes x, we can proceed to use Minitab as follows.

Minitab Procedure for Obtaining Confidence Intervals for p

1. Select **Stat** from the main menu.

2. Select **Basic Statisticss**.

3. Select **1 Proportion**.

4. You will get a dialog box like the one shown below. The entries correspond to $n = 1501$ and $x = 1051$.

5. Click on **Options** to change the confidence level from 95% to any other value. (The **Options** button can also be used to check the box for "Test and interval based on normal distribution." You could click on that box to use the same procedure given in the textbook, but it is better to use the *exact* procedure, so leave that box unchecked.)

After clicking **OK**, the Minitab results are as shown below.

Test and CI for One Proportion

```
Sample       X     N  Sample p              95% CI
1         1051  1501  0.700200   (0.676313, 0.723301)
```

Based on the above Minitab display, we can express the 95% confidence interval estimate of p as $0.676313 < p < 0.723301$. After rounding, the confidence interval becomes $0.676 < p < 0.723$. Example 3 in Section 7-2 of the textbook shows that when the normal approximation method is used, the confidence interval is $0.677 < p < 0.723$, so the exact method and the normal approximation method produce results that are very close in this case.

The textbook stresses that the ultimate objective is not to simply construct the confidence interval; the ultimate objective is to *interpret* and understand the confidence interval. Example 3 in the textbook includes this question: Based on the results, can we safely conclude that the majority of adults believe in global warming? The solution in Example 3 includes the statement that based on the confidence interval, it does appear that the proportion of adults who believe in global warming is greater than 0.5 (or 50%), so we can safely conclude that the majority of adults believe in global warming. Because the limits of 0.677 and 0.723 are likely to contain the true population proportion, it appears that the population proportion is a value greater than 0.5.

7-2 Confidence Intervals for Estimating μ

Sections 7-3 and 7-4 in the textbook introduce confidence intervals used for estimating the value of a population mean μ. The textbook stresses the importance of selecting the correct distribution (normal or t). The textbook provides a summary of the criteria, and the table below is included.

Choosing between z and t

Method	Conditions
Use normal (z) distribution.	σ **known and normally distributed population** *or* σ **known and $n > 30$**
Use t distribution.	σ **not known and normally distributed population** *or* σ **not known and $n > 30$**
Use a nonparametric method or bootstrapping.	**Population is not normally distributed and $n \leq 30$**

The textbook also notes that in choosing between the normal or t distributions, a major criterion for choosing between the normal (z) or Student t distributions is whether the population standard deviation σ is known. In reality, it is very rare to know σ, so the t distribution is typically used instead of the normal distribution.

Here is the procedure for using Minitab to construct confidence interval estimates of a population mean μ.

Minitab Procedure for Obtaining Confidence Intervals for μ

1.   Select **Stat** from the main menu.

2.   Select **Basic Statistics** from the subdirectory.

3.   Select either **1-Sample** z or **1-Sample t** by using the criteria summarized in the preceding table.

4.   You will now see a dialog box, such as the one shown below. For the sample data, you can use either the original list of sample values by selecting **Samples in** columns, or you can use the summary statistics by selecting **Summarized** data. (If you select 1-Sample z in Step 3, there will also be a box for entering "Sigma," the population standard deviation σ.) Enter column label or the summary statistics.

5.   Click on the **Options** box to enter the confidence level. (The Minitab default is a confidence level of 95%.) Click **OK** when done.

If you click **OK** on the above screen, you get the following Minitab display.

One-Sample T

N	Mean	StDev	SE Mean	95% CI
49	0.40	21.00	3.00	(-5.63, 6.43)

The above screen display includes entries corresponding to Example 2 in Section 7-4 from the textbook, which is shown below in an abbreviated form. Note that **1-Sample t** was selected in Step 3 (because σ is not known and $n > 30$).

> **Garlic for Reducing Cholesterol** A common claim is that garlic lowers cholesterol levels. In a test of the effectiveness of garlic, 49 subjects were treated with doses of raw garlic, and their cholesterol levels were measured before and after the treatment. The changes in their levels of LDL cholesterol (in mg/dL) have a mean of 0.4 and a standard deviation of 21.0 (based on data from "Effect of Raw Garlic vs Commercial Garlic Supplements on Plasma Lipid Concentrations in Adults With Moderate Hypercholesterolemia," by Gardner et al, *Archives of Internal Medicine*, Vol. 167). Use the sample statistics to construct a 95% confidence interval estimate of the mean net change in LDL cholesterol after the garlic treatment. What does the confidence interval suggest about the effectiveness of garlic in reducing LDL cholesterol?

Based on the preceding Minitab display, the 95% confidence interval is $-5.6 < \mu < 6.4$ (rounded). The solution in the textbook includes this: On the basis of the given sample results, we are 95% confident that the limits of -5.6 and 6.4 actually do contain the value of μ, the mean of the changes in LDL cholesterol for the population. Because the confidence interval limits contain the value of 0, it is very possible that the mean of the changes in LDL cholesterol is equal to 0, suggesting that the garlic treatment did not affect the LDL cholesterol levels. It appears that the garlic treatment is *not effective* in lowering LDL cholesterol.

7-3 Confidence Intervals for Estimating σ

Here is the Minitab procedure for constructing confidence interval estimates of the population standard deviation σ. (If you want a confidence interval for the variance σ^2, use the following procedure and square the resulting confidence interval limits.) Note that this procedure requires a list of the original sample data.

Minitab Procedure for Confidence Intervals for Estimating σ

1. Enter the original sample values in a Minitab column.

2. Select **Stat** from the main menu.

3. Select **Basic Statistics** from the subdirectory.

4. Select **1 Variance**.

5. You will get a dialog box that allows you to enter the column containing the sample data or the summary statistics. Either enter the column (such as C1) or enter the summary statistics.

6. Click **OK**.

Consider Example 2 from Section 7-5 in the textbook:

> **Confidence Interval for Home Voltage** The proper operation of typical home appliances requires voltage levels that do not vary much. Listed below are ten voltage levels (in volts) recorded in the author's home on ten different days. (The voltages are from Data Set 13 in Appendix B.) These ten values have a standard deviation of $s = 0.15$ volt. Use the sample data to construct a 95% confidence interval estimate of the standard deviation of all voltage levels.
>
> 123.3 123.5 123.7 123.4 123.6 123.5 123.5 123.4 123.6 123.8

If you use the Minitab procedure with the 10 voltages listed in this example, you will obtain the Minitab display on the following page. The relevant results are highlighted with boldface type, and we can see that the 95% confidence interval is $0.10 < \sigma < 0.27$ (rounded). Also, the 95% confidence interval limit for the variance is $0.01 < \sigma^2 < 0.07$ (rounded).

Test and CI for One Standard Deviation: C1

```
Method
The standard method is only for the normal distribution.
The adjusted method is for any continuous distribution.


Statistics
Variable   N  StDev  Variance
C1        10  0.149    0.0223


95% Confidence Intervals
                                          CI for
Variable  Method      CI for StDev       Variance
C1        Standard    (0.103, 0.273)   (0.0106, 0.0744)
          Adjusted    (0.104, 0.266)   (0.0108, 0.0705)
```

7-4 Determining Sample Sizes

The textbook discuss procedures for determining sample sizes necessary to estimate a population proportion p, or mean μ, or standard deviation σ. Minitab does not have the ability to calculate sample sizes directly, but determinations of sample size use simple formulas, so it is easy to use a calculator with the procedures described in the textbook. Minitab could be used instead of a calculator. Consider this formula for determining the sample size required to estimate a population mean:

$$n = \left[\frac{z_{\alpha/2}\sigma}{E}\right]^2$$

Suppose that we want to find the sample size needed to estimate the mean IQ score of statistics professors, and we want 95% confidence that the error in the sample mean is no more than 2 IQ points. We further assume that $\sigma = 15$. The required value of z can be found by using the procedure described in Section 6-1 of this manual/workbook. For example, to find the z score corresponding to 95% in two tails, select **Inverse cumulative probability** and enter a cumulative left area of 0.975 to get the z score of 1.95996. Then enter the Minitab command LET C2=(1.95996*15/2)**2 to complete the calculation for the sample size. The result will be the sample size of 216.082, which we round up to 217. The determination of sample size for estimating a population proportion p can be handled the same way. The determination of sample size for estimating σ or σ^2 can be done by using Table 7-2 in the textbook.

CHAPTER 7 EXPERIMENTS: Confidence Intervals and Sample Sizes

In Experiments 7–1 through 7–4, use Minitab with the sample data and confidence level to construct the confidence interval estimate of the population proportion p.

7–1. $n = 400, x = 300$, 95% confidence   _____

7–2. $n = 1200, x = 200$, 99% confidence   _____

7–3. $n = 1655, x = 176$, 98% confidence   _____

7–4. $n = 2001, x = 1776$, 90% confidence   _____

In Experiments 7–5 through 7–12, use Minitab with the sample data and confidence level to construct the confidence interval estimate of the population mean μ. Be careful to correctly choose between the t and normal distributions.

7–5. $n = 45, \bar{x} = 75, s = 12$, 95% confidence   _____

7–6. $n = 45, \bar{x} = 75, \sigma = 12$, 95% confidence   _____

7–7. $n = 32, \bar{x} = 102, \sigma = 14$, 99% confidence   _____

7–8. $n = 32, \bar{x} = 102, s = 14$, 99% confidence   _____

7–9. $n = 81, \bar{x} = 12.8, s = 2.7$, 95% confidence   _____

7–10. $n = 76, \bar{x} = 1008.1, \sigma = 47.3$, 95% confidence   _____

7–11. $n = 118, \bar{x} = 2.75, \sigma = 0.33$, 95% confidence   _____

7–12. $n = 76, \bar{x} = 86.7, s = 10.8$, 95% confidence   _____

7–13. *Mendelian Genetics* When Mendel conducted his famous genetics experiments with peas, one sample of offspring consisted of 428 green peas and 152 yellow peas.

   a.   Use Minitab to find the following confidence interval estimates of the percentage of yellow peas.
   99.5% confidence interval: _____
   99% confidence interval: _____
   98% confidence interval: _____
   95% confidence interval: _____
   90% confidence interval: _____

(continued)

b.   After examining the pattern of the above confidence intervals, complete the following statement. "As the degree of confidence decreases, the confidence interval limits

_____."

c.   In your own words, explain why the preceding completed statement makes sense. That is, why should the confidence intervals behave as you have described?

_____

_____

7-14.   *Gender Selection* The Genetics and IVF Institute conducted a clinical trial of the XSORT method designed to increase the probability of conceiving a girl. As of this writing, 574 babies were born to parents using the XSORT method, and 525 of them were girls.

a.   What is the best point estimate of the population proportion of girls born to parents using the XSORT method? _____

b.   Use the sample data to construct a 95% confidence interval estimate of the percentage of girls born to parents using the XSORT method.

_____

c.   Based on the results, does the XSORT method appear to be effective? Why or why not?

_____

_____

7-15.   *Gender Selection* The Genetics and IVF Institute conducted a clinical trial of the YSORT method designed to increase the probability of conceiving a boy. As of this writing, 152 babies were born to parents using the YSORT method, and 127 of them were boys.

a.   What is the best point estimate of the population proportion of boys born to parents using the YSORT method? _____

b.   Use the sample data to construct a 99% confidence interval estimate of the percentage of boys born to parents using the YSORT method.

_____

c.   Based on the results, does the YSORT method appear to be effective? Why or why not?

_____

7-16. ***Touch Therapy*** When she was nine years of age, Emily Rosa did a science fair experiment in which she tested professional touch therapists to see if they could sense her energy field. She flipped a coin to select either her right hand or her left hand, then she asked the therapists to identify the selected hand by placing their hand just under Emily's hand without seeing it and without touching it. Among 280 trials, the touch therapists were correct 123 times (based on data in "A Close Look at Therapeutic Touch," *Journal of the American Medical Association*, Vol. 279, No. 13).

    a.   Given that Emily used a coin toss to select either her right hand or her left hand, what proportion of correct responses would be expected if the touch therapists made random guesses? _____

    b.   Using Emily's sample results, what is the best point estimate of the therapist's success rate?_____

    c.   Using Emily's sample results, construct a 99% confidence interval estimate of the proportion of correct responses made by touch therapists.

_____

    d.   What do the results suggest about the ability of touch therapists to select the correct hand by sensing an energy field?

_____

In Exercises 7-17 through 7-20, use the data sets from Appendix B in the textbook.

7-17. ***Nicotine in Cigarettes*** Refer to Data Set 4 in Appendix B of the textbook and assume that the samples are simple random samples obtained from normally distributed populations. The Minitab worksheet is CIGARET.

    a.   Construct a 95% confidence interval estimate of the mean amount of nicotine in cigarettes that are king size, non-filtered, non-menthol, and non-light.

_____

    b.   Construct a 95% confidence interval estimate of the mean amount of nicotine in cigarettes that are 100 mm, filtered, non-menthol, and non-light.

_____

    c.   Compare the results. Do filters on cigarettes appear to be effective?

_____

_____

7-18. ***Pulse Rates*** A physician wants to develop criteria for determining whether a patient's pulse rate is atypical, and she wants to determine whether there are significant differences between males and females. Use the sample pulse rates in Data Set 1 from Appendix B in the textbook. The Minitab worksheets are MHEALTH and FHEALTH.

a.   Construct a 95% confidence interval estimate of the mean pulse rate for males.

_____

b.   Construct a 95% confidence interval estimate of the mean pulse rate for females.

_____

c.   Compare the preceding results. Can we conclude that the population means for males and females are different? Why or why not?

_____

_____

7-19. ***Boston Rainfall*** Refer to the Monday and Saturday rainfall amounts listed in Data Set 14 in Appendix B of the textbook. The Minitab worksheet is BOSTRAIN.

a.   Construct a 95% confidence interval estimate of the mean amount of rainfall on Monday.

_____

b.   Construct a 95% confidence interval estimate of the mean amount of rainfall on Saturday.

_____

c.   Compare the preceding results. Can we conclude that the population means for Monday and Saturday are different? Why or why not?

_____

_____

7-20. ***Weights of Coins*** Refer to the weights of pre-1964 quarters and the weights of post-1964 quarters listed in Data Set 20 from Appendix B of the textbook. The Minitab worksheet is COINS.

    a.   Construct a 95% confidence interval estimate of the mean weight of pre-1964 quarters.

_____

    b.   Construct a 95% confidence interval estimate of the mean weight of post-1964 quarters.

_____

    c.   Compare the preceding results. Can we conclude that the population means for pre-1964 quarters and post-1964 quarters are different? Why or why not?

_____

_____

7-21. ***Simulated Data*** Minitab is designed to generate random numbers from a variety of different sampling distributions. In this experiment we will generate 500 IQ scores, then we will construct a confidence interval based on the sample results. IQ scores have a normal distribution with a mean of 100 and a standard deviation of 15. First generate the 500 sample values as follows.

    1.   Click on **Calc,** then select **Random Data**, then **Normal**.

    2.   In the dialog box, enter a sample size of 500, a mean of 100, and a standard deviation of 15. Specify column C1 as the location for storing the results. Click **OK**.

    3.   Use **Stat/Basic Statistics/Display Descriptive Statistics** to find these statistics:

        $n =$ _____ $\bar{x} =$ _____ $s =$ _____

Using the generated values, construct a 95% confidence interval estimate of the population mean of all IQ scores. Enter the 95% confidence interval here.

_____

Because of the way that the sample data were generated, we *know* that the population mean is 100. Do the confidence interval limits contain the true mean IQ score of 100?

_____

(continued)

If this experiment were to be repeated over and over again, how often would we expect the confidence interval limits to contain the true population mean value of 100? Explain how you arrived at your answer.

_____

_____

7-22.   ***Simulated Data***   Follow the same steps listed in Experiment 7-21 to randomly generate 500 IQ scores from a population having a normal distribution, a mean of 100, and a standard deviation of 15. Record the sample statistics here.

$n =$ _____ $\bar{x} =$ _____ $s =$ _____

Confidence intervals are typically constructed with confidence levels around 90%, 95%, or 99%. Instead of constructing such a typical confidence interval, use the generated values to construct a 50% confidence interval. Enter the result below.

_____

Does the above confidence interval have limits that actually do contain the true population mean, which we know is 100? _____

Repeat the above procedure 9 more times and list the resulting 50% confidence intervals here.

_____   _____

_____   _____

_____   _____

_____   _____

_____

Among the total of the 10 confidence intervals constructed, how many of them actually do contain the true population mean of 100? Is this result consistent with the fact that the level of confidence used is 50%? Explain.

_____

_____

7-23.   ***Combining Data Sets***   Refer to the M&M Data Set 18 in Appendix B of the textbook and use the entire sample of 100 plain M&M candies to construct a 95% confidence interval for the mean weight of all M&Ms. The Minitab worksheet is M&M. (*Hint*: It is not necessary to manually enter the 100 weights, because they are already stored in separate Minitab columns in the worksheet M&M. Use **Stack** to combine the different M&M data sets into one big data set. Now construct a 95% confidence interval estimate of the population mean of all IQ scores. Enter the 95% confidence interval here.

_____

7-24.   ***Combining Data Sets***   Refer to the word counts in Data Set 8 in Appendix B of the textbook. The Minitab worksheet is WORDS. Combine all of the word counts from males in one column, and combine all of the word counts from females in another column.

   a.   Find the 95% confidence interval estimate of the mean word count for men.

   _____

   b.   Find the 95% confidence interval estimate of the mean word count for women.

   _____

   c.   Based on the results, does it appear the women talk more than men? Why or why not?

   _____

   _____

In Exercises 7-25 through 7-28, use the data sets from Appendix B in the textbook to construct confidence interval estimates of the population standard deviations.

7-25.   ***Nicotine in Cigarettes***   Refer to Data Set 4 in Appendix B of the textbook and assume that the samples are simple random samples obtained from normally distributed populations. The Minitab worksheet is CIGARET.

   a.   Construct a 95% confidence interval estimate of the standard deviation of amounts of nicotine in cigarettes that are king size, non-filtered, non-menthol, and non-light. _____

   b.   Construct a 95% confidence interval estimate of the standard deviation of amounts of nicotine in cigarettes that are 100 mm, filtered, non-menthol, and non-light. _____

   c.   Compare the results. Do the amounts of variation appear to be different?

   _____

7-26.   ***Pulse Rates*** A physician wants to develop criteria for determining whether a patient's pulse rate is atypical, and she wants to determine whether there are significant differences between males and females. Use the sample pulse rates in Data Set 1 from Appendix B in the textbook. The Minitab worksheets are MHEALTH and FHEALTH.

   a.   Construct a 95% confidence interval estimate of the standard deviation of pulse rates for males.

   _____

   b.   Construct a 95% confidence interval estimate of the standard deviation of pulse rates for females.

   _____

   c.   Compare the preceding results. Do the amounts of variation appear to be different?

   _____

   _____

7-27.   ***Boston Rainfall*** Refer to the Monday and Saturday rainfall amounts listed in Data Set 14 in Appendix B of the textbook. The Minitab worksheet is BOSTRAIN.

   a.   Construct a 95% confidence interval estimate of the standard deviation of the amounts of rainfall on Monday.

   _____

   b.   Construct a 95% confidence interval estimate of the standard deviation of the amounts of rainfall on Saturday.

   _____

   c.   Compare the preceding results. Can we conclude that the amounts of variation are different?

   _____

   _____

7-28. ***Weights of Coins*** Refer to the weights of pre-1964 quarters and the weights of post-1964 quarters listed in Data Set 20 from Appendix B of the textbook. The Minitab worksheet is COINS.

   a. Construct a 95% confidence interval estimate of the standard deviation of the weights of pre-1964 quarters.

_____

   b. Construct a 95% confidence interval estimate of the standard deviation of the weights of post-1964 quarters.

_____

   c. Compare the preceding results. Do the amounts of variation appear to be different?

_____

_____

8

Hypothesis Testing

8–1 Testing Hypotheses About p

Important note about methods: The textbook gives a method for testing a claim about a population proportion p. The textbook procedure is based on using a normal distribution as an approximation to a binomial distribution. However, *Minitab uses an* exact *calculation instead of a normal approximation.* (Minitab provides an option for using the normal approximation method, but it is better to use the exact procedure.)

> ## Minitab uses an *exact* procedure (instead of using a normal approximation) for testing hypothesis about a population proportion p.

The Minitab procedure for testing claims about a population proportion follows the same basic steps described in Section 7–1 of this manual/workbook. In addition to having a claim to be tested, Minitab also requires a significance level, the sample size n, and the number of successes x.

 Finding x: In Section 7-1 of this manual/workbook, we briefly discussed one particular difficulty that arises when the available information provides the sample size n and the sample proportion $\hat{p}$ instead of the number of successes x. We provided this procedure for determining the number of successes x.

> **To find the number of successes x from the sample proportion and sample size:**
> **Calculate $x = \hat{p}\,n$ and round the result to the nearest whole number.**

Given a claim about a proportion and knowing n and x, we can use Minitab as follows.

Minitab Procedure for Testing Claims about p

 1. Select **Stat** from the main menu.

 2. Select **Basic Statistics** from the subdirectory.

 3. Select **1 Proportion**.

 4. You will get a dialog box.

- Click on the option for summarized data.
- In the box for "Number of trials," enter the value of the sample size n.
- In the box for the number of events (or successes), enter the number of successes x.
- Click on the **Options** bar and enter the confidence level. (Enter 95 for a 0.05 significance level.) Also, in the "Test proportion" box, enter the claimed value of the population proportion p. Also select the format of the alternative hypothesis.
- Click **OK**.

For example, let's use the sample data of $n = 726$ and $x = 4668$ to test the claim that $p > 0.5$. (See Example 1 in Section 8-3 of the textbook.) The dialog boxes and Minitab results are as follows.

Based on the above dialog boxes, the following Minitab results are obtained.

Test and CI for One Proportion

```
Test of p = 0.5 vs p > 0.5
```

Sample	X	N	Sample p	95% Lower Bound	Exact P-Value
1	668	726	0.920110	0.901566	0.000

We can see from this display that the *P*-value is 0.000, which causes us to reject the null hypothesis and support the claim that $p > 0.5$. (See Example 1 in Section 8-3 of the textbook.)

8–2 Testing Hypotheses About μ

The textbook describes procedures for testing claims about the mean of a single population. When testing claims about the mean of a single population, the textbook explains that there are different procedures, depending on the size of the sample, the nature of the population distribution, and whether the population standard deviation σ is known. The textbook stresses the importance of selecting the correct distribution (normal or t). See the following table, which summarizes the criteria for choosing between the normal and t distributions.

Choosing between z and t

Method	Conditions
Use normal (z) distribution.	**σ known and normally distributed population** *or* **σ known and $n > 30$**
Use t distribution.	**σ not known and normally distributed population** *or* **σ not known and $n > 30$**
Use a nonparametric method or bootstrapping.	**Population is not normally distributed and $n \le 30$**

Minitab allows us to test claims about a population mean μ by using either the list of original sample values or the summary statistics of n, $\overline{x}$, and s.

Minitab Procedure for Testing Claims About μ

1. Select **Stat** from the main menu.

2. Select **Basic Statistics** from the subdirectory.

3. Select either **1-Sample z** or **1-Sample t** by using the criteria summarized in the above table. (If the methods of this chapter do not apply, you're stuck. You must use other methods, such as bootstrap resampling or nonparametric methods.)

4. You will now see a dialog box, such as the one shown below. If you select 1-Sample z in Step 4, there will also be a box for entering "Sigma," the population standard deviation σ.

5. In the "Samples in columns" box, enter the column containing the list of sample data, or if the summary statistics are known, click **Summarized data** and proceed to enter the sample size, sample mean, and sample standard deviation.

6. Click on the box labeled "Perform hypothesis test."

7.   In the "Hypothesized mean" box, enter the claimed value of the population mean.

8.   Click on the **Options** button to get the options dialog box, and make the entries according to the following.

- Enter a confidence level. (Enter 95 for a hypothesis test conducted with a 0.05 significance level.)

- Click on the "Alternative" box to change the default alternative hypothesis. The available choices in the "Alternative" box are
    less than
    not equal
    greater than

- Click **OK** when done.

9.   Click **OK** on the main dialog box.

As an illustration, consider this, from Example 1 in Section 8-5:

> **Overloading Boats** Using the weights of the simple random sample of men from Data Set 1 in Appendix B, we obtain these sample statistics: $n = 40$ and $\bar{x} = 172.55$ lb, and $s = 26.32$ lb. Do not assume that the value of σ is known. Use these results to test the claim that men have a mean weight greater than 166.3 lb. This is, text the claim that $\mu > 166.3$ lb. Use a 0.05 significance level

Because σ is not known and $n > 30$, we use Minitab's **1-Sample t** (instead of 1-Sample z). The following dialog boxes correspond to the above example.

We get these Minitab results:

One-Sample T

```
Test of mu = 166.3 vs > 166.3
```

```
                          95% Lower
  N    Mean   StDev  SE Mean      Bound      T       P
 40  172.55   26.32     4.16     165.54   1.50   0.071
```

From the above Minitab display, we see that the test statistic is $t = 1.50$. The P-value is 0.071. Because the P-value is greater than the significance level of 0.05, we fail to reject the null hypothesis and conclude that there is not sufficient evidence to support the claim that the population mean is greater than 166.3 lb. (See Example 1 in Section 8-5 in the textbook.)

8-3 Testing Hypotheses About σ or σ^2

When testing claims about σ or σ^2, the textbook stresses that there are the following two important requirements:

1.   The samples are simple random samples. (Remember the importance of good sampling methods.)

2.   The sample values come from a population with a *normal distribution*.

The textbook makes the very important point that *for tests of claims about standard deviations or variances, the requirement of a normal distribution is very strict.* If the population does not have a normal distribution, then inferences about standard deviations or variances can be very misleading. *Suggestion:* Given sample data, assess normality by using the methods from Section 6-4 of this manual/workbook. If the population distribution does appear to have a normal distribution and you want to test a claim about the population standard deviation or variance, use the Minitab procedure given below. (The **1 Variance** option is not available on Minitab Release 14 and earlier versions.)

Minitab Procedure for Testing Claims about σ or σ^2

1.   Select **Stat** from the main menu.

2.   Select **Basic Statistics** from the subdirectory.

3.   Select **1 Variance**.

4.   In the top box, select the option of **Enter standard deviation**.

5.   In the "Samples in columns" box, enter the column containing the list of sample data, or if the summary statistics are known, click **Summarized data** and proceed to enter the sample size and sample standard deviation.

6.   Click on the box labeled "Perform hypothesis test."

7.   In the "Hypothesized standard deviation" box, enter the claimed value of the population standard deviation.

8.   Click on the **Options** button to get the options dialog box, and make the entries according to the following.

- Enter a confidence level. (Enter 95 for a hypothesis test conducted with a 0.05 significance level.)

- Click on the "Alternative" box to change the default alternative hypothesis. The available choices in the "Alternative" box are

  less than
  not equal
  greater than

- Click **OK** when done.

9.   Click **OK** on the main dialog box.

As an illustration, consider this from Example 1 in Section 8-6:

> **Quality Control of Coins** Consider the simple random sample of the 37 weights of post-1983 pennies listed in Data Set 20 in Appendix B. Those 37 weights have a standard deviation of 0.01648 g. Use a 0.05 significance level to test the claim that the population of weights has a *standard deviation* less than the specification of 0.0230 g.

From the above example, we see that we want to test the claim that $\sigma < 0.0230$ g, and we want to use a 0.05 significance level. Sample data consist of the summary statistics $n = 37$ and $s = 0.01648$ g. See Example 1 in Section 8-6 of the textbook, where the normality of the distribution is considered with a Minitab-generated histogram and a STATDISK-generated normal quantile plot. There are no outliers and the sample does appear to be from a population having a normal distribution, as required. We can proceed with the hypothesis test, and the Minitab results are shown below. Because the *P*-value of 0.007 is less than the significance level of 0.05, we reject the null hypothesis and conclude that there is sufficient sample evidence to support the claim that the population has a standard deviation less than 0.0230 g.

Test and CI for One Standard Deviation

```
Method      Chi-Square  DF   P-Value
Standard        18.48   36     0.007
```

The following page shows the Minitab screens that produce the above results.

1 Variance   ☒

| Enter standard deviation   ▼ |

○ Samples in columns

    C1   ▲
                                                                             ▼

◉ Summarized data

    Sample size: 37

    Sample standard deviation: 0.01648

☑ Perform hypothesis test

    Hypothesized standard deviation: 0.0230

    Select   Options...

    Help   OK   Cancel

1 Variance - Options   ☒

Confidence level: 95.0

Alternative: less than   ▼

    Help   OK   Cancel

8-4   Testing Hypotheses with Simulations

Sections 8-1, 8-2, and 8-3 of this manual/workbook have all focused on using Minitab for hypothesis tests using the *P*-value approach or the confidence interval approach. Another very different approach is to use *simulations*. Let's illustrate the simulation technique with an example.

Consider testing the claim that the population of body temperatures of healthy adults has a mean less than 98.6°F. That is, consider the claim that $\mu < 98.6$°F. We will use sample data consisting of these 12 values:

$$98.0\ \ 97.5\ \ 98.6\ \ 98.8\ \ 98.0\ \ 98.5\ \ 98.6\ \ 99.4\ \ 98.4\ \ 98.7\ \ 98.6\ \ 97.6$$

For these 12 values, $n = 12$, $\bar{x} = 98.39167$, and $s = 0.53506$. The key question is this:

> If the population mean body temperature is really 98.6, then *how likely* is it that we would get a sample mean of 98.39167, given that the population has a normal distribution and the sample size is 12?

If the probability of getting a sample mean such as 98.39167 is very small, then that suggests that the sample results are not the result of chance random fluctuation. If the probability is high, then we can accept random chance as an explanation for the discrepancy between the sample mean of 98.39167 and the assumed mean of 98.6. What we need is some way of determining the likelihood of getting a sample mean such as 98.39167. That is the precise role of *P*-values in the *P*-value approach to hypothesis testing. However, there is another approach. Minitab and many other software packages are capable of generating random results from a variety of different populations. Here is how Minitab can be used: Determine the likelihood of getting a sample mean of 98.39167 by randomly generating different samples from a population that is normally distributed with the claimed mean of 98.6. For the standard deviation, we will use the best available information: the value of $s = 0.53506$ obtained from the sample.

Minitab Procedure for Testing Hypotheses with Simulations

1.   Identify the values of the sample size *n*, the sample standard deviation *s*, and the claimed value of the population mean.

2.   Click on **Calc**, then select **Random Data**.

3.   Click on **Normal.**

4.   Generate a sample randomly selected from a population with the claimed mean. When making the required entries in the dialog box, use the *claimed* mean, the sample size *n*, and the sample standard deviation *s*.

5. Continue to generate similar samples until it becomes clear that the given sample mean is or is not likely. (Here is one criterion: The given sample mean is *unlikely* if its value or more extreme values occur 5% of the time or less.) If it is unlikely, reject the claimed mean. If it is likely, fail to reject the claimed mean.

For example, here are 20 results obtained from the random generation of samples of size 12 from a normally distributed population with a mean of 98.6 and a standard deviation of 0.53506:

98.754	**98.374**	**98.332**	98.638	98.513
98.551	98.566	98.760	**98.332**	98.603
98.407	98.640	98.655	98.408	98.802
98.505	98.699	98.609	98.206	98.582

Examining the 20 sample means, we see that three of them are 98.39167 or lower. Because 3 of the 20 results (or 15%) are at least as extreme as the sample mean of 98.39167, we see that a sample mean such as 98.39167 is *common* for these circumstances. This suggests that a sample mean of 98.39167 is not *significantly* different from the assumed mean of 98.6. We would feel more confident in this conclusion if we had more sample results, so we could continue to randomly generate simulated samples until we feel quite confident in our thinking that a sample mean such as 98.39167 is not an unusual result. It can easily occur as the result of chance random variation. We therefore fail to reject the null hypothesis that the mean equals 98.6. There is not sufficient evidence to support the claim that the mean is less than 98.6.

CHAPTER 8 EXPERIMENTS: Hypothesis Testing

Experiments 8–1 through 8–4 involve claims about proportions. Use Minitab's exact procedure to test the given claim.

8–1. ***Reporting Income*** In a Pew Research Center poll of 745 randomly selected adults, 589 said that it is morally wrong to not report all income on tax returns. Use a 0.01 significance level to test the claim that 75% of adults say that it is morally wrong to not report all income on tax returns.

Test statistic: _____   *P*–value: _____

Conclusion in your own words: _____

_____

8–2. ***Voting for the Winner*** In a presidential election, 308 out of 611 voters surveyed said that they voted for the candidate who won (based on data from ICR Survey Research Group). Use a 0.01 significance level to test the claim that among all voters, the percentage who believe that they voted for the winning candidate is equal to 43%, which is the actual percentage of votes for the winning candidate.

Test statistic: _____   *P*–value: _____

Conclusion in your own words: _____

_____

8–3. ***Tennis Instant Replay*** The Hawk-Eye electronic system is used in tennis for displaying an instant replay that shows whether a ball is in bounds or out of bounds. In the first U.S. Open that used the Hawk-Eye system, players could challenge calls made by referees. The Hawk-Eye system was then used to confirm or overturn the referee's call. Players made 839 challenges, and 327 of those challenges were successful with the call overturned (based on data reported in *USA Today*). Use a 0.01 significance level to test the claim that the proportion of challenges that are successful is greater than 1/3.

Test statistic: _____   *P*–value: _____

Conclusion in your own words: _____

_____

8–4.   ***Screening for Marijuana Usage***   The company Drug Test Success provides a "1-Panel-THC" test for marijuana usage. Among 300 tested subjects, results from 27 subjects were wrong (either a false positive or a false negative). Use a 0.05 significance level to test the claim that less than 10% of the test results are wrong.

Test statistic: _____   *P*–value: _____

Conclusion in your own words: _____

_____

8-5.   ***Exact Procedure and Normal Approximation***   Repeat Experiment 8-4 by using Minitab with the normal approximation procedure. (Click on **Options** and click on the box to **Use test and interval based on normal distribution**.)

Test statistic: _____   *P*–value: _____

Conclusion in your own words: _____

_____

Compare the exact results obtained in Experiment 8-4 to those found by using the normal approximation.

_____

_____

Experiments 8–6 through 8–9 use Minitab with the original list of sample data for testing the given claim about a population mean.

8–6.   ***Do the Screws Have a Length of 3/4 in.?***   A simple random sample of 50 stainless steel sheet metal screws is obtained from those suppled by Crown Bolt, Inc., and the length of each screw is measured using a vernier caliper. The lengths are listed in Data Set 19 of Appendix B. The Minitab data set is SCREWS. Use a 0.05 significance level to test the claim that the screws have a mean length equal to 3/4 in. (or 0.75 in.), as indicated on the package labels.

Test statistic: _____   *P*–value: _____

Conclusion in your own words: _____

_____

8–7. ***Power Supply*** Data Set 13 in Appendix B lists measured voltage amounts supplied directly to the author's home. The Minitab data set is VOLTAGE. The Central Hudson power supply company states that it has a target power supply of 120 volts. Using those home voltage amounts, test the claim that the mean is 120 volts. Use a 0.01 significance level.

Test statistic: _____   *P*–value: _____

Conclusion in your own words: _____

_____

8–8. ***Is 98.6°F Wrong?*** Data Set 2 in Appendix B includes measured human body temperatures. The Minitab data set is BODYTEMP. Use the temperatures listed for 12 AM on day 2 to test the common belief that the mean body temperature is 98.6°F. Does that common belief appear to be wrong?

Test statistic: _____   *P*–value: _____

Conclusion in your own words: _____

_____

8-9. ***FICO Credit Scores*** Data Set Set 24 in Appendix B includes a simple random sample of FICO credit rating scores. The Minitab data set is FICO. As of this writing, the mean FICO score was reported to be 678. Use a 0.05 significance level to test the claim that the sample of FICO scores come from a population with a mean equal to 678.

Test statistic: _____   *P*–value: _____

Conclusion in your own words: _____

_____

In Experiments 8–10 through 8–13, test the claim about a population mean.

8–10. ***Cigarette Tar*** A simple random sample of 25 filtered 100 mm cigarettes is obtained, and the tar content of each cigarette is measured. The sample has a mean of 13.2 mg and a standard deviation of 3.7 mg (based on Data Set 4 in Appendix B). Use a 0.05 significance level to test the claim that the mean tar content of filtered 100 mm cigarettes is less than 21.1 mg, which is the mean for unfiltered king size cigarettes.

Test statistic: _____   *P*–value: _____

Conclusion in your own words: _____

_____

8–11. ***Is the Diet Practical?*** When 40 people used the Weight Watchers diet for one year, their mean weight *loss* was 3.0 lb and the standard deviation was 4.9 lb (based on data from "Comparison of the Atkins, Ornish, Weight Watchers, and Zone Diets for Weight Loss and Heart Disease Reduction," by Dansinger, et al., *Journal of the American Medical Association,* Vol. 293, No. 1). Use a 0.01 significance level to test the claim that the mean weight loss is greater than 0 lb.

Test statistic: _____   *P*–value: _____

Conclusion in your own words: _____

_____

Based on these results, does the diet appear to be effective? Does the diet appear to have practical significance? Explain.

_____

_____

8–12. ***Weights of Pennies*** The U. S. Mint has a specification that pennies have a mean weight of 2.5 g. Data Set 20 in Appendix B lists the weights (in grams) of 37 pennies manufactured after 1983. Those pennies have a mean weight of 2.49910 g and a standard deviation of 0.01648 g. Use a 0.05 significance level to test the claim that this sample is from a population with a mean weight equal to 2.5 g.

Test statistic: _____   *P*–value: _____

Conclusion in your own words: _____

_____

8–13. ***Analysis of Pennies*** In an analysis investigating the usefulness of pennies, the cents portions of 100 randomly selected credit card charges are recorded. The sample has a mean of 47.6 cents and a standard deviation of 33.5 cents. (The amounts are listed in Table 3-1.) If the amounts from 0 cents to 99 cents are all equally likely, the mean is expected to be 49.5 cents. Use a 0.01 significance level to test the claim that the sample is from a population with a mean equal to 49.5 cents.

Test statistic: _____   *P*–value: _____

Conclusion in your own words: _____

In Experiments 8–14 through 8–17, use Minitab to test the claim about a standard deviation or variance.

8–14. **BMI for Miss America** Listed below are body mass indexes (BMI) for recent Miss America winners. Use a 0.01 significance level to test the claim that recent Miss America winners are from a population with a standard deviation of 1.34, which was the standard deviation of BMI for winners from the 1920s and 1930s. Do recent winners appear to have variation that is different from that of the 1920s and 1930s?

         19.5   20.3   19.6   20.2   17.8   17.9   19.1   18.8   17.6   16.8

Test statistic: _____   *P*–value: _____

Conclusion in your own words: _____

_____

8–15. **Vitamin Supplement and Birth Weight** Listed below are birth weights (in kilograms) of male babies born to mothers on a special vitamin supplement (based on data from the New York State Department of Health). Test the claim that this sample comes from a population with a standard deviation equal to 0.470 kg, which is the standard deviation for male birth weights in general. Does the vitamin supplement appear to affect the variation among birth weights?

         3.73   4.37   3.73   4.33   3.39   3.68   4.68   3.52
         3.02   4.09   2.47   4.13   4.47   3.22   3.43   2.54

Test statistic: _____   *P*–value: _____

Conclusion in your own words: _____

_____

8-16. **Aircraft Altimeters** The Skytek Avionics company uses a new production method to manufacture aircraft altimeters. A simple random sample of new altimeters resulted in errors listed below. Use a 0.05 level of significance to test the claim that the new production method has errors with a standard deviation greater than 32.2 ft, which was the standard deviation for the old production method.

         −42   78   −22   −72   −45   15   17   51   −5   −53   −9   −109

Test statistic: _____   *P*–value: _____

Conclusion in your own words: _____

_____

8–17. ***Playing Times of Popular Songs*** Listed below are the playing times (in seconds) of songs that were popular at the time of this writing. (The songs are by Timberlake, Furtado, Daughtry, Stefani, Fergie, Akon, Ludacris, Beyonce, Nickelback, Rihanna, Fray, Lavigne, Pink, Mims, Mumidee, and Omarion.) Use a 0.05 significance level to test the claim that the songs are from a population with a standard deviation less than one minute.

448 242 231 246 246 293 280 227 244 213 262 239 213 258 255 257

Test statistic: _____   *P*–value: _____

Conclusion in your own words: _____

_____

The following experiments involve the simulation approach to hypothesis testing.

8–18. ***Hypothesis Testing with Simulations*** Suppose that we want to use a 0.05 significance level to test the claim that $p > 0.5$ and we have sample data summarized by $n = 100$ and $x = 70$. Use the following Minitab procedure for using a simulation method for testing the claim that $p > 0.5$.

We assume in the null hypothesis that $p = 0.5$. Working under the assumption that $p = 0.5$, we will generate different samples of size $n = 100$ until we have a sense for the likelihood of getting 70 or more successes. Begin by generating a sample of size 100 as follows: Select **Calc/Random Data/Integer**. In the dialog box, enter 100 for the number of rows, enter C1–C10 for the columns, then enter a minimum of 0 and a maximum of 1. After clicking **OK**, the result will be 10 columns, where each column represents a sample of 100 subjects with a probability of 0.5 of getting a 1 in each cell. Letting 1 represent a success, use **Stat/Display Descriptive Statistics** and find the number of samples having a sample proportion of 0.70 or greater. Enter that result here: _____

What percentage of the simulated samples have a sample proportion of 0.70 or greater? Enter that result here: _____

What does the preceding result suggest about the likelihood of this event: "When the population proportion *p* is really equal to 0.5, a sample of size 100 is randomly selected and the sample proportion is 0.70 or greater."

_____

Can we conclude that the probability of a sample proportion of 0.70 or greater is less than or equal to the significance level of 0.05? _____

What is the final conclusion? _____

_____

8–19. *Hypothesis Testing with Simulations* Experiment 8–18 used a simulation approach to test a claim. Use a simulation approach for conducting the hypothesis test described in Experiment 8–4. Describe the procedure, results, and conclusions.

_____

_____

_____

_____

8–20. *Hypothesis Testing with Simulations* Refer to Experiment 8–10 in this manual/work-book. We want to test the claim that $\mu < 21.1$, and the given sample data can be summarized with these statistics: $n = 25$, $\bar{x} = 13.2$, and $s = 3.7$. Instead of conducting a formal hypothesis test, we will now consider another way of determining whether the sample mean of 13.2 is significantly less than the claimed value of 21.1. We will use a significance level of 0.05 with this criterion:

The sample mean of 13.2 is significantly less than 21.1 if there is a 5% chance (or less) that the following event occurs: A sample mean of 13.2 or less is obtained when selecting a random sample of $n = 25$ values from a normally distributed population with mean $\mu = 21.1$ and standard deviation $\sigma = 3.7$.

Use **Calc/Random Data/Normal** to randomly generate a sample of 25 values from a normally distributed population with the assumed mean of 21.1 and a standard deviation of 3.7. Find the mean of the generated sample and record it in the space below. Then generate another sample, and another sample, and continue until you have enough sample means to determine how often a result such as $\bar{x} = 13.2$ (or less) will occur.

_____

_____

Enter the number of generated samples:_____

How many of the generated samples have a sample mean less than 21.1? _____

What is the proportion of trials in which the sample mean was 21.1 or less? _____

Based on these results, what do you conclude about the claim that $\mu < 21.1$? Explain.

_____

9

Inferences from Two Samples

9-1 Two Proportions

The textbook makes the point that the section discussing inferences involving two proportions is one of the most important sections in the book because the main objective is to provide methods for dealing with two sample proportions — a situation that is very common in real applications.

To use Minitab, we must first identify the number of successes x_1 and the sample size n_1 for the first sample, and identify x_2 and n_2 for the second sample. Sample data often consist of sample proportions or percentages instead of the actual numbers of successes, so we must know how to determine the number of successes. From $\hat{p}_1 = x_1/n_1$, we know that $x_1 = n_1 \cdot \hat{p}_1$ so that x_1 can be found by multiplying the sample size for the first sample by the sample proportion expressed in decimal form. For example, if 35.2% of 1200 interviewed subjects answer "yes" to a question, we have $n = 1200$ and $x = 1200 \cdot 0.352 = 422.4$, which we round to 422.

After identifying n_1, x_1, n_2, and x_2, proceed with Minitab as follows.

Minitab Procedure for Inferences (Hypothesis Tests and Confidence Intervals)
for Two Proportions

To conduct hypothesis tests or construct confidence intervals for two population proportions, use the following procedure.

1.   For each sample, find the sample size n and the number of successes x.

2.   Select **Stat**, then **Basic Statistics**, then **2 Proportions**.

3.   In the dialog box, select the option of "Summarized data" and enter the number of successes (or "events") for each of the two samples, and enter the sample size (or "trials") for each of the two samples. (The numbers of "Events" are the values of x_1 and x_2. The numbers of "Trials" are the values of n_1 and n_2.)

4.   Click on the **Options** bar and proceed to enter the confidence level (enter 95 for a 0.05 significance level). If conducting a hypothesis test, enter the claimed value of the difference (usually 0), and select the format for the alternative hypothesis. (See the Minitab screen on the next page where "greater than" is selected for the format of the alternative hyothesis.)

5.   While still in the Options dialog box, refer to the box with the label of "Use pooled estimate of p for test."

     - If testing a hypothesis, click on that box (because you want to use a pooled estimate of p).

     - If constructing a confidence interval, do *not* click on that box (because you do not want to use a pooled estimate of p).

As an illustration, consider this example:

> **Do airbags save lives?** The table below lists results from a simple random
> sample of front-seat occupants involved in car crashes (based on data from
> "Who Wants Airbags?" by Meyer and Finney, *Chance*, Vol. 18, No. 2).
> Use a 0.05 significance level to test the claim that the fatality rate of
> occupants is lower for those in cars equipped with airbags.

	Airbag Available	No Airbag Available
Occupant Fatalities	41	52
Total number of occupants	11,541	9,853

For notation purposes, we stipulate that sample 1 is the group with airbags available, and
sample 2 is the group with no airbags available. We can now proceed to use the above Minitab
procedure. After selecting **Stat**, **Basic Statistics**, then **2 Proportions**, we make the required
entries in the dialog boxes as shown below. The claim of a lower fatality rate for the population
with airbags available is the claim that $p_1 < p_2$, so we select the option of "less than" for the
alternative hypothesis.

2 Proportions (Test and Confid... ✕

  ○ Samples in one column

    Samples:

    Subscripts:

  ○ Samples in different columns

    First:

    Second:

  ● Summarized data

	Events:	Trials:
First:	41	11541
Second:	52	9853

Select

Options...

Help   OK   Cancel

---

2 Proportions - Options ✕

Confidence level: 95.0

Test difference: 0.0

Alternative: less than ▼

☑ Use pooled estimate of p for test

Help   OK   Cancel

After clicking on **OK** in both dialog boxes, the Minitab results will be displayed. Shown below are the results from the preceding example and dialog box entries. The results include the test statistic and *P*–value. The display shows a *P*–value of 0.028, which is less than the significance level of 0.05, so we reject the null hypothesis and support the claim that the fatality rate is lower for the population with airbags available.

Test and CI for Two Proportions

```
Sample   X       N  Sample p
1       41   11541  0.003553
2       52    9853  0.005278


Difference = p (1) - p (2)
Estimate for difference:  -0.00172503
95% upper bound for difference:  -0.000217916
Test for difference = 0 (vs < 0):   Z = -1.91  P-Value = 0.028


Fisher's exact test: P-Value = 0.036
```

Confidence Intervals: The above Minitab display includes a one–sided confidence interval, but it is based on the pooling of the two sample proportions. (See Step 5 in the preceding Minitab procedure. Note that we check the box for pooling the sample proportions only if we are testing a hypothesis. If we are constructing a confidence interval, we should *not* check the box for pooling the sample proportions.) If we use the given sample data for constructing a 95% confidence interval estimate of $p_1 - p_2$, we get $-0.00323 < p_1 - p_2 < -0.0000218$ from the results shown below. Remember, when clicking on the Options box, select "not equal to" for the alternative hypothesis, and do *not* click on the box for pooling the sample standard deviations.

Test and CI for Two Proportions

```
Sample   X       N  Sample p
1       41   11541  0.003553
2       52    9853  0.005278


Difference = p (1) - p (2)
Estimate for difference:  -0.00172503
90% CI for difference:  (-0.00323214, -0.000217916)
Test for difference = 0 (vs not = 0):   Z = -1.88   P-Value = 0.060

Fisher's exact test: P-Value = 0.061
```

9-2 Two Means: Independent Samples

The textbook notes that two samples are **independent** if the sample values selected from one population are not related to or somehow paired or matched with the sample values selected from the other population. If there is some relationship so that each value in one sample is paired with a corresponding value in the other sample, the samples are **dependent.** Dependent samples are often referred to as **matched pairs,** or **paired samples.**

　　　When testing a claim about the means of two independent samples, or when constructing a confidence interval estimate of the difference between the means of two independent samples, the textbook describes procedures based on the requirement that the two population standard deviations σ_1 and σ_2 are not known, and there is no assumption that $\sigma_1 = \sigma_2$. We focus on the first of the following three cases, and the other two cases are discussed briefly:

1. σ_1 and σ_2 are not known and are not assumed to be equal.
2. σ_1 and σ_2 are known.
3. It is assumed that $\sigma_1 = \sigma_2$.

The first of these three cases is the main focus of Section 9-3 in the textbook, and Minitab handles that first case but, instead of using "the smaller of $n_1 - 1$ and $n_2 - 1$" for the number of degrees of freedom, Minitab calculates the number of degrees of freedom using Formula 9-1 in *Elementary Statistics,* 11th edition.

Minitab also provides the option of assuming that $\sigma_1 = \sigma_2$, but there is no option for assuming that σ_1 and σ_2 are *known*. That is, Minitab allows us to deal with cases 1 and 3 listed above, but not case 2. We describe the Minitab procedure, then consider an example.

Minitab Procedure for Inferences (Hypothesis Tests and Confidence Intervals) about Two Means: Independent Samples

Use the following procedure for both hypothesis tests and confidence intervals.

1. Identify the values of the summary statistics, or enter the original sample values in columns C1 and C2 (or open a Minitab worksheet with columns containing the lists of sample values).

2. Click on the main menu item of **Stat**, select **Basic Statistics**, then **2-Sample t**.

3. Proceed to make the entries in the dialog box.

   - Do *not* check the box labeled "Assume equal variances" (unless there is a sound justification for assuming that $\sigma_1 = \sigma_2$). Be sure that the box remains unchecked. We are not assuming equal variances.

   - Click on the **Options** bar and enter the confidence level (enter 95 for a 0.05 significance level). If conducting a hypothesis test, enter the claimed value of the difference (usually 0), and also select the format for the alternative hypothesis. Click **OK** when finished.

As an illustration, consider the example on the following page.

Are Men and Women Equal Talkers? A headline in *USA Today* proclaimed that "Men, women are equal talkers." That headline referred to a study of the numbers of words that samples of men and women spoke in a day. Given below are the results from the study, which are included in Data Set 8 in Appendix B (based on "Are Women Really More Talkative Than Men?" by Mehl, et al, *Science*, Vol. 317, No. 5834). Use a 0.05 significance level to test the claim that men and women speak the same mean number of words in a day. Does there appear to be a difference?

Number of Words Spoken in a Day

Men	Women
$n_1 = 186$	$n_2 = 210$
$\bar{x}_1 = 15{,}668.5$	$\bar{x}_2 = 16{,}215.0$
$s_1 = 8632.5$	$s_2 = 7301.2$

Using Minitab with the preceding procedure, we enter the summary statistics as shown below.

Click on the **Options** button, then make the selections shown in the following dialog box.

2-Sample t - Options

Confidence level: 95.0

Test difference: 0.0

Alternative: not equal ▼

Help   OK   Cancel

The Minitab results are as shown below. The test statistic is $t = -0.68$ and the P–value is 0.500, so we fail to reject the null hypothesis. There is not sufficient evidence to reject the claim that men and women speak the same number of words in a day.

Two-Sample T-Test and CI

```
Sample    N    Mean   StDev  SE Mean
1        186  15669   8633      633
2        210  16215   7301      504

Difference = mu (1) - mu (2)
Estimate for difference:  -547
95% CI for difference:  (-2137, 1044)
T-Test of difference = 0 (vs not =): T-Value = -0.68   P-Value = 0.500
DF = 364
```

Confidence Interval If we wanted to construct a confidence interval for the difference between the mean number of words spoken by men and the mean number of words spoken by women in day, the results above show that the 95% confidence interval is $-2137 < \mu_1 - \mu_2 < 1044$. Because that confidence interval includes 0, there does not appear to be a significant difference between the two means.

9-3 Matched Pairs

The textbook describes methods for testing hypotheses and constructing confidence interval estimates of the differences between samples consisting of *matched pairs*. Here is the Minitab procedure:

Minitab Procedure for Inferences (Hypothesis Tests and Confidence Intervals) about Two Means: Matched Pairs

To conduct hypothesis tests or construct confidence intervals using data consisting of matched pairs, use the following procedure.

1. Enter the values of the first sample in column C1.
    Enter the values of the second sample in column C2
    (or open a Minitab worksheet that includes columns of matched pairs of sample data).

2. Select **Stat**, **Basic Statistics**, **Paired t**.

3. Make these entries and selections in the dialog box that pops up:

    - Enter C1 (or the column label) in the box labeled "First sample.
    - Enter C2 (or the column label) in the box labeled "Second sample."
    - Click the **Options** bar and proceed to enter the confidence level (enter 95 for a 0.05 significance level). If testing a claim about the mean of the differences between the paired data, enter the claimed value of the difference μ_d, and select the form of the alternative hypothesis. Click **OK**, then click **OK** for the main dialog box.

See the following example.

> **Hypothesis Test of Claimed Freshman Weight Gain** Listed below are samples of measured weights of college students in September and April of their freshman year. Use the sample data with a 0.05 significance level to test the claim that for the population of students, the mean change in weight from September to April is equal to 0 kg.

Weight (kg) Measurements of Students in Their Freshman Year					
April weight	66	52	68	69	71
September weight	67	53	64	71	70

Shown on the next page are the dialog boxes and the Minitab results.

Paired T-Test and CI: April, September

```
Paired T for April - September

            N    Mean   StDev   SE Mean
April       5   65.20    7.60      3.40
September   5   65.00    7.25      3.24
Difference  5    0.20    2.39      1.07

95% CI for mean difference: (-2.76, 3.16)
T-Test of mean difference = 0 (vs not = 0): T-Value = 0.19  P-Value = 0.861
```

Based on the above results, the hypothesis test yields a *P*-value of 0.861, which is greater than the significance level of 0.05, so we fail to reject the null hypothesis. We fail to reject the claim that the mean of the differences between the April weights and the September weights is equal to 0 kg.

Confidence Interval The preceding Minitab results show that the 95% confidence interval estimate of the mean of the differences is −2.8 kg < μ_d < 3.2 kg (rounded). Because the confidence interval includes 0 kg, there does not appear to be a significant difference.

9-4 Two Variances

The textbook describes the use of the F distribution in testing a claim that two populations have the same variance (or standard deviation). Minitab is capable of conducting such an F test.

Minitab Procedure for Inferences (Hypothesis Tests and Confidence Intervals) for Two Standard Deviations or Two Variances

1. Either obtain the summary statistics for each sample, or enter the individual sample values in two columns (or open a Minitab worksheet containing the columns of sample data).

2. Select **Stat**, then **Basic Statistics**, then **2 Variances**.

3. A dialog box will appear. Either select the option of "Samples in different columns" and proceed to enter the column names or labels, or select "Summarized data" and proceed to enter the summary statistics. (If entering the summary statistics, be sure to enter the sample *variances*, which can be obtained by squaring the sample standard deviations.)

4. Click on the **Options** bar and enter the confidence level. (Enter 95 for a 0.05 significance level.) Click **OK**, then click **OK** in the main dialog box.

Consider the following example, which is followed by the Minitab results obtained by using the above procedure.

> **Comparing Variation in Weights of Quarters** Sample statistics are listed below for weights of quarters made before and after 1964. Use a 0.05 significance level to test the claim that the weights of pre-1964 quarters and the weights of post-1964 quarters are from populations with the same standard deviation.
>
> | Pre-1964 Quarters | Post-1964 Quarters |
> |---|---|
> | $n = 40$ | $n = 40$ |
> | $s = 0.08700$ g | $s = 0.06194$ g |

Test for Equal Variances

```
F-Test (Normal Distribution)
Test statistic = 1.97, p-value = 0.037
```

The Minitab results include the P-value of 0.037, which is less than the significance level of 0.05. We therefore reject the null hypothesis that the two populations of quarters have weights with the same standard deviation.

CHAPTER 9 EXPERIMENTS: Inferences from Two Samples

9-1.   ***Drug Use in College*** In a 1993 survey of 560 college students, 171 said that they used illegal drugs during the previous year. In a recent survey of 720 college students, 263 said that they used illegal drugs during the previous year (based on data from the National Center for Addiction and Substance Abuse at Columbia University). Use a 0.05 significance level to test the claim that the proportion of college students using illegal drugs in 1993 was less than it is now.

Test statistic: _____   *P*–value: _____

Conclusion in your own words: _____

_____

9-2.   ***Drug Use in College*** Using the sample data from Experiment 9-1, construct the confidence interval corresponding to the hypothesis test conducted with a 0.05 significance level. What conclusion does the confidence interval suggest?

_____

_____

9-3.   ***Are Seat Belts Effective?*** A simple random sample of front-seat occupants involved in car crashes is obtained. Among 2823 occupants not wearing seatbelts, 31 were killed. Among 7765 occupants wearing seatbelts, 16 were killed (based on data from "Who Wants Airbags?" by Meyer and Finney, *Chance*, Vol. 18, No. 2). Construct a 90% confidence interval estimate of the difference between the fatality rates for those not wearing seat belts and those wearing seat belts. What does the result suggest about the effectiveness of seat belts?

_____

_____

9-4.   ***Are Seat Belts Effective?*** Use the sample data in Experiment 9-3 with a 0.05 significance level to test the claim that the fatality rate is higher for those not wearing seat belts.

Test statistic: _____   *P*–value: _____

Conclusion in your own words: _____

_____

9-5. ***Morality and Marriage*** A Pew Research Center poll asked randomly selected subjects if they agreed with the statement that "It is morally wrong for married people to have an affair." Among the 386 women surveyed, 347 agreed with the statement. Among the 359 men surveyed, 305 agreed with the statement. Use a 0.05 significance level to test the claim that the percentage of women who agree is different from the percentage of men who agree. Does there appear to be a difference in the way women and men feel about this issue?

Test statistic: _____   *P*–value: _____

Conclusion in your own words: _____

_____

9-6. ***Morality and Marriage*** Using the sample data from Experiment 9-5, construct the confidence interval corresponding to the hypothesis test conducted with a 0.05 significance level. What conclusion does the confidence interval suggest?

_____

_____

9-7. ***Raising the Roof in Baseball*** In a recent baseball World Series, the Houston Astros were ordered to keep the roof of the Astrodome open. The Houston team claimed that this would make them lose a home-field advantage, because the noise from fans would be less effective. During the regular season, Houston won 36 of 53 games played with the roof closed, and they won 15 of 26 games played with the roof open. Treat these results as simple random samples, and use a 0.05 significance level to test the claim that the proportion of wins at home is higher with a closed roof than with an open roof. Does the closed roof appear to be an advantage?

Test statistic: _____   *P*–value: _____

Conclusion in your own words: _____

_____

9-8. ***Raising the Roof in Baseball*** Using the sample data from Experiment 9-7, construct the confidence interval corresponding to the hypothesis test conducted with a 0.05 significance level. What conclusion does the confidence interval suggest?

_____

_____

In Experiments 9–9 through 9–20, assume that the two samples are independent simple random samples selected from normally distributed populations. Do not assume that the population standard deviations are equal.

9-9. ***Hypothesis Test of Effectiveness of Humidity in Treating Croup*** Treatment In a randomized controlled trial conducted with children suffering from viral croup, 46 children were treated with low humidity while 46 other children were treated with high humidity. Researchers used the Westley Croup Score to assess the results after one hour. The low humidity group had a mean score of 0.98 with a standard deviation of 1.22 while the high humidity group had a mean score of 1.09 with a standard deviation of 1.11 (based on data from "Controlled Delivery of High vs Low Humidity vs Mist Therapy for Croup Emergency Departments," by Scolnik et al, *Journal of the American Medical Association,* Vol. 295, No. 11). Use a 0.05 significance level to test the claim that the two groups are from populations with the same mean. What does the result suggest about the common treatment of humidity?

Test statistic: _____   *P*–value: _____

Conclusion in your own words: _____

_____

9-10. ***Confidence Interval for Effectiveness of Humidity in Treating Croup*** Refer to the sample data given in Experiment 9-9 and construct a 95% confidence interval estimate of the difference between the mean Westley Croup Score of children treated with low humidity and the mean score of children treated with high humidity. What does the confidence interval suggest about humidity as a treatment for croup?

_____

_____

9-11. ***Confidence Interval for Cigarette Tar*** The mean tar content of a simple random sample of 25 unfiltered king-size cigarettes is 21.1 mg, with a standard deviation of 3.2 mg. The mean tar content of a simple random sample of 25 filtered 100 mm cigarettes is 13.2 mg with a standard deviation of 3.7 mg (based on data from Data Set 4 in Appendix B). Construct a 90% confidence interval estimate of the difference between the mean tar content of unfiltered king size cigarettcs and the mean tar content of filtered 10mm cigarettes. Does the result suggest that 100 mm filtered cigarettes have less tar than unfiltered king size cigarettes?

_____

_____

9-12. *Hypothesis Test for Cigarette Tar* Refer to the sample data in Experiment 9-11 and use a 0.05 significance level to test the claim that unfiltered king size cigarettes have a mean tar content greater than that of filtered 100 mm cigarettes. What does the result suggest about the effectiveness of cigarette filters?

Test statistic: _____   *P*–value: _____

Conclusion in your own words: _____

_____

9-13. *BMI for Miss America* The trend of thinner Miss America winners has generated charges that the contest encourages unhealthy diet habits among young women. Listed below are body mass indexes (BMI) for Miss America winners from two different time periods. Consider the listed values to be simple random samples selected from larger populations.

BMI (from recent winners):	19.5 20.3 19.6 20.2 17.8 17.9 19.1
	18.8 17.6 16.8
BMI (from the 1920's and 1930's):	20.4 21.9 22.1 22.3 20.3 18.8 18.9
	19.4 18.4 19.1

a. Use a 0.05 significance level to test the claim that recent winners have a lower mean BMI than winners from the 1920's and 1930's.

Test statistic: _____   *P*–value: _____

Conclusion in your own words: _____

_____

b. Construct a 90% confidence interval for the difference between the mean BMI of recent winners and the mean BMI of winners from the 1920's and 1930's.

_____

_____

9-14. **_Radiation in Baby Teeth_** Listed below are amounts of Strontium-90 (in millibecquerels or mBq per gram of calcium) in a simple random sample of baby teeth obtained from Pennsylvania residents and New York residents born after 1979 (based on data from "An Unexpected Rise in Strontium-90 in U.S. Deciduous Teeth in the 1990s," by Mangano, et. al., _Science of the Total Environment_).

Pennsylvania: 155 142 149 130 151 163 151 142 156 133 138 161

New York: 133 140 142 131 134 129 128 140 140 140 137 143

a.   Use a 0.05 significance level to test the claim that the mean amount of Strontium-90 from Pennsylvania residents is greater than the mean amount from New York residents.

Test statistic: _____   _P_–value: _____

Conclusion in your own words: _____

_____

b.   Construct a 90% confidence interval of the difference between the mean amount of Strontium-90 from Pennsylvania residents and the mean amount from New York residents.

_____

_____

9-15. **_Longevity_** Listed below are the numbers of years that popes and British monarchs (since 1690) lived after their election or coronation (based on data from _Computer-Interactive Data Analysis_, by Lunn and McNeil, John Wiley & Sons). Treat the values as simple random samples from a larger population.

Popes: 2   9   21   3   6   10   18   11   6   25   23   6   2   15   32   25
                11   8   17   19   5   15   0   26

Kings and Queens: 17   6   13   12   13   33   59   10   7   63   9   25   36   15

a.   Use a 0.01 significance level to test the claim that the mean longevity for popes is less the mean for British monarchs after coronation.

Test statistic: _____   _P_–value: _____

Conclusion in your own words: _____

_____

(_continued_)

b.   Construct a 98% confidence interval of the difference between the mean longevity of popes and the mean longevity for kings and queens. What does the result suggest about those two means?

_____

_____

9-16. ***Sex and Blood Cell Counts***   White blood cell counts are helpful for assessing liver disease, radiation, bone marrow failure, and infectious diseases. Listed below are white blood cell counts found in simple random samples of males and females (based on data from the Third National Health and Nutrition Examination Survey).

Female: 8.90 6.50 9.45 7.65 6.40 5.15 16.60 5.75 11.60 5.90 9.30 8.55
             10.80 4.85 4.90 8.75 6.90 9.75 4.05 9.05 5.05 6.40 4.05 7.60
             4.95 3.00 9.10
Male: 5.25 5.95 10.05 5.45 5.30 5.55 6.85 6.65 6.30 6.40 7.85 7.70 5.30
             6.50 4.55 7.10 8.00 4.70 4.40 4.90 10.75 11.00 9.60

a.   Use a 0.01 significance level to test the claim that females and males have different mean white blood cell counts.

Test statistic: _____ P–value: _____

Conclusion in your own words: _____

_____

b.   Construct a 98% confidence interval of the difference between the mean white blood cell count of females and males. Based on the result, does there appear to be a difference?

_____

_____

9-17. ***Appendix B Data Set: Movie Income***   Refer to Data Set 9 in Appendix B. Use the amounts of money grossed by movies with ratings of PG or PG-13 as one sample, and use the amounts of money grossed by movies with R ratings.
a.   Use a 0.01 significance level to test the claim that movies with ratings of PG or PG-13 have a higher mean gross amount than movies with R ratings.

Test statistic: _____ P–value: _____

Conclusion in your own words: _____

_____

(continued)

b.   Construct a 98% confidence interval estimate of the difference between the mean amount of money grossed by movies with ratings of PG or PG-13 and the mean amount of money grossed by movies with R ratings. What does the confidence interval suggest about movies as an investment?

_____

_____

9-18   *Appendix B Data Set: Word Counts* Refer to Data Set 8 in Appendix B. Use the word counts for male and female psychology students recruited in Mexico (see the columns labeled 3M and 3F).

a.   Use a 0.05 significance level to test the claim that male and female psychology students speak the same mean number of words in a day.

Test statistic: _____   *P*–value: _____

Conclusion in your own words: _____

_____

b.   Construct a 95% confidence interval estimate of the difference between the mean number of words spoken in a day by male and female psychology students in Mexico. Do the confidence interval limits include 0, and what does that suggest about the two means?

_____

_____

9-19.   *Appendix B Data Set: Voltage* Refer to Data Set 13 in Appendix B. Use a 0.05 significance level to test the claim that the sample of home voltages and the sample of generator voltages are from populations with the same mean. If there is a statistically significant difference, does that difference have practical significance?

Test statistic: _____   *P*–value: _____

Conclusion in your own words: _____

_____

9-20. ***Appendix B Data Set: Weights of Coke*** Refer to Data Set 17 in Appendix B and test the claim that because they contain the same amount of cola, the mean weight of cola in cans of regular Coke is the same as the mean weight of cola in cans of Diet Coke. If there is a difference in the mean weights, identify the most likely explanation for that difference.

Test statistic: _____   *P*–value: _____

Conclusion in your own words: _____

_____

9-21. ***Does BMI Change During Freshman Year?*** Listed below are body mass indices (BMI) of students. The BMI of each student was measured in September and April of the freshman year (based on data from "Changes in Body Weight and Fat Mass of Men and Women in the First Year of College: A Study of the 'Freshman 15'," by Hoffman, Policastro, Quick, and Lee, *Journal of American College Health*, Vol. 55, No. 1). Use a 0.05 significance level to test the claim that the mean change in BMI for all students is equal to 0. Does BMI appear to change during freshman year?

April BMI	20.15	19.24	20.77	23.85	21.32
September BMI	20.68	19.48	19.59	24.57	20.96

Test statistic: _____   *P*–value: _____

Conclusion in your own words: _____

_____

9-22. ***Confidence Interval for BMI Changes*** Use the same paired data from Experiment 9-21 to construct a 95% confidence interval estimate of the change in BMI during freshman year. Does the confidence interval include 0, and what does that suggest about BMI during freshman year?

_____

_____

9-23. ***Are Best Actresses Younger than Best Actors?*** Listed below are ages of actresses and actors at the times that they won Oscars. The data are paired according to the years that they won. Use a 0.05 significance level to test the common belief that best actresses are younger than best actors. Does the result suggest a problem in our culture?

Best Actresses   28 32 27 27 26 24 25 29 41 40 27 42 33 21 35
Best Actors   62 41 52 41 34 40 56 41 39 49 48 56 42 62 29

Test statistic: _____   *P*–value: _____

Conclusion in your own words: _____

_____

9-24. ***Are Flights Cheaper When Scheduled Earlier?*** Listed below are the costs (in dollars) of flights from New York (JFK) to San Francisco for US Air, Continental, Delta, United, American, Alaska, and Northwest. Use a 0.01 significance level to test the claim that flights scheduled one day in advance cost more than flights scheduled 30 days in advance. What strategy appears to be effective in saving money when flying?

Flight scheduled one day in advance 456 614 628 1088 943 567 536
Flight scheduled 30 days in advance 244 260 264   264 278 318 280

Test statistic: _____   *P*–value: _____

Conclusion in your own words: _____

_____

9-25. ***Does Your Body Temperature Change During the Day?*** Listed below are body temperatures (in °F) of subjects measured at 8 AM and at 12 AM (from University of Maryland physicians listed in Data Set 2 in Appendix B). Construct a 95% confidence interval estimate of the difference between the 8 AM temperatures and the 12 AM temperatures. Is body temperature basically the same at both times?

| 8 AM | 97.0 | 96.2 | 97.6 | 96.4 | 97.8 | 99.2 |
| 12 AM | 98.0 | 98.6 | 98.8 | 98.0 | 98.6 | 97.6 |

Test statistic: _____   *P*–value: _____

Conclusion in your own words: _____

_____

9-26. ***Is Blood Pressure the Same for Both Arms?*** Listed below are systolic blood pressure measurements (mm Hg) taken from the right and left arms of the same woman (based on data from "Consistency of Blood Pressure Differences Between the Left and Right Arms," by Eguchi et al, *Archives of Internal Medicine*, Vol. 167). Use a 0.05 significance level to test for a difference between the measurements from the two arms. What do you conclude?

Right arm	102	101	94	79	79
Left arm	175	169	182	146	144

Test statistic: _____   *P*–value: _____

Conclusion in your own words: _____

_____

9-27. ***Appendix B Data Set: Paper or Plastic?*** Refer to Data Set 22 in Appendix B. Construct a 95% confidence interval estimate of the mean of the differences between weights of discarded paper and weights of discarded plastic. Which seems to weigh more: discarded paper or discarded plastic?

_____

_____

9-28. ***Appendix B Data Set: Glass and Food*** Refer to Data Set 22 in Appendix B. Construct a 95% confidence interval estimate of the mean of the differences between weights of discarded glass and weights of discarded food. Which seems to weigh more: discarded glass or discarded food?

_____

_____

9-29. ***Baseline Characteristics*** In journal articles about clinical experiments, it is common to include *baseline characteristics* of the different treatment groups so that they can be compared. In an article about the effects of different diets, a table of baseline characteristics showed that 40 subjects treated with the Atkins diet had a mean age of 47 years with a standard deviation of 12 years. Also, 40 subjects treated with the Zone diet had a mean age of 51 years with a standard deviation of 9 years. Use a 0.05 significance level to test the claim that subjects from both treatment groups have ages with the same amount of variation.

Test statistic: _____   *P*–value: _____

Conclusion in your own words: _____

_____

9-30. ***Braking Distances of Cars*** A random sample of 13 four-cylinder cars is obtained, and the braking distances are measured and found to have a mean of 137.5 ft and a standard deviation of 5.8 ft. A random sample of 12 six-cylinder cars is obtained and the braking distances have a mean of 136.3 ft and a standard deviation of 9.7 ft (based on Data Set 16 in Appendix B). Use a 0.05 significance level to test the claim that braking distances of four-cylinder cars and braking distances of six-cylinder cars have the same standard deviation.

Test statistic: _____ *P*–value: _____

Conclusion in your own words: _____

_____

9-31. ***Freshman 15 Study*** Use the sample weights (kg) of male and female college students measured in April of their freshman year, as listed in Data Set 3 in Appendix B. Use a 0.05 significance level to test the claim that near the end of the freshman year, weights of male college students vary more than weights of female college students.

Test statistic: _____ *P*–value: _____

Conclusion in your own words: _____

_____

9-32. ***Heights*** Use the samples of heights of men and women listed in Data Set 1 in Appendix B and use a 0.05 significance level to test the claim that heights of men vary more than heights of women.

Test statistic: _____ *P*–value: _____

Conclusion in your own words: _____

_____

10

Correlation and Regression

10-1 Scatterplot

Chapter 10 in the Triola textbook introduces the basic concepts of linear correlation and regression. The basic objective is to use paired sample data to determine whether there is a relationship between two variables and, if so, identify what the relationship is. Consider the paired sample data in the table below. The data consist of the cost of a slice of pizza and the price of a subway fare in New York City. The data are paired by year. We want to determine whether there is a relationship between the cost of a slice of pizza and the price of a subway fare. If such a relationship exists, we want to identify it with an equation so that we can predict the price of a subway fare based on the cost of a slice of pizza.

Cost of a Slice of Pizza, Subway Fare, and the CPI

Cost of Pizza	0.15	0.35	1.00	1.25	1.75	2.00
Subway Fare	0.15	0.35	1.00	1.35	1.50	2.00

Before using concepts of correlation and regression, it is always wise to construct a scatterplot so that we can see if there appears to be a relationship between the two variables, and we can see if the pattern of points is approximately the pattern of a straight line. The Minitab procedure for constructing a scatterplot was given in Section 2-8 of this manual/workbook. Enter the paired data in columns of Minitab or open a Minitab worksheet, select **Graph**, then **Scatterplot**. Select the option of **Simple**, then click **OK**.

10-2 Correlation

The textbook describes the linear correlation coefficient r as a measure of the strength of the linear relationship between two variables. Minitab can compute the value of r for paired data. Instead of doing the complicated manual calculations, follow these steps to let Minitab calculate r.

Minitab Procedure for Calculating the Linear Correlation Coefficient r

1. Enter the x values in column C1, and enter the corresponding y values in column C2 (or open a Minitab worksheet containing the columns of paired data). It is also wise to enter the names of the variables in the empty cells located immediately above the first row of data values.

2. Select **Stat**, then **Basic Statistics**, then **Correlation**.

3. In the dialog box, enter the column labels of C1 C2 (or enter the variable names), then click **OK**.

Shown below is the Minitab display that results from using the data in the above table. The variable names of Pizza and Subway were also entered. Because the P-value of 0.000 is less than the significance level of 0.05, we conclude that there is a significant linear correlation between the cost of a slice of pizza and the price of a subway fare.

Correlations: Pizza, Subway

```
Pearson correlation of Pizza and Subway = 0.988
P-Value = 0.000
```

10-3 Regression

The textbook discusses the important topic of regression. Given a collection of paired sample data, the **regression equation**

$$\hat{y} = b_0 + b_1x$$

describes the relationship between the two variables. The graph of the regression equation is called the **regression line** (or *line of best fit*, or *least-squares line*). The regression equation expresses a relationship between x (called the *independent variable* or *predictor variable*) and y (called the *dependent variable* or *response variable*). The y-intercept of the line is b_0 and the slope is b_1. Given a collection of paired data, we can find the regression equation as follows.

Minitab Procedure for Finding the Equation of the Regression Line

1.   Enter the x values in column C1, and enter the corresponding y values in column C2 (or open a Minitab worksheet containing the columns of paired data). It is also wise to enter the names of the variables in the empty cells located immediately above the first row of data values.

2.   Select **Stat**, then **Regression**, then **Regression** once again.

3.   Make these entries in the dialog box:

- Enter C2 (the y values) in the box for the "Response" variable.

- Enter C1 (the x values) in the box for the "Predictor" variable.

- Click on the **Options** bar and click on the "**Fit intercept**" box.

- Click on the **Results** bar and select the second option that includes the regression equation, s, and R-Squared.

- Click **OK**.

Using the sample data in the preceding table, if you use the response variable of Subway and the predictor variable of Pizza, the Minitab display will be as follows.

Regression Analysis: Subway versus Pizza

```
The regression equation is
Subway = 0.0346 + 0.945 Pizza


Predictor     Coef   SE Coef       T      P
Constant   0.03456   0.09501    0.36  0.734
Pizza      0.94502   0.07446   12.69  0.000


S = 0.122987   R-Sq = 97.6%   R-Sq(adj) = 97.0%


Analysis of Variance


Source          DF      SS      MS        F       P
Regression       1  2.4366  2.4366   161.09   0.000
Residual Error   4  0.0605  0.0151
Total            5  2.4971
```

This display includes the regression equation, which we express as $\hat{y} = 0.0346 + 0.945x$. The Minitab display also includes "R-sq = 97.6%," which is the coefficient of determination defined in the textbook; that value indicates that 97.6% of the variation in prices of subway fares can be explained by the costs of a slice of pizza, and 2.4% of the variation in prices of a subway fare remains unexplained. The display also includes other items discussed later in the textbook.

In addition to the *equation* of the regression line, we can also get a *graph* of the regression line superimposed on the scatter diagram.

Minitab Procedure for Graph of Regression Line

1.   Enter the *x* values in column C1, and enter the corresponding *y* values in column C2 (or open a Minitab worksheet containing the columns of paired data). It is also wise to enter the names of the variables in the empty cells located immediately above the first row of data values.

2.   Select **Stat**, **Regression**, then **Fitted line plot**.

3.   Complete the dialog box by entering the response (*y*) variable, the predictor (*x*) variable, and select the **Linear** model. Click **OK**.

Shown below is the graph of the regression line for the data in the preceding table. We can see that the regression line fits the points quite well.

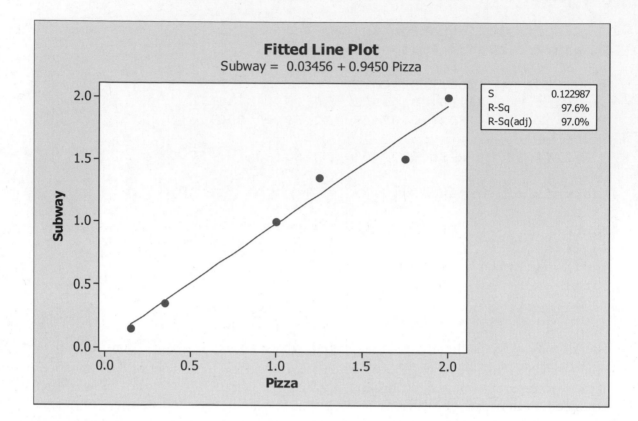

10-4 Predictions

We can use the regression equation for making predictions. For example, suppose that we want to predict the price of a subway fare when the cost of a slice of pizza is $2.25. To find the predicted value of y given $x = 2.25$, follow the same procedure for finding the equation of the regression line as given in Section 10-3 of this manual/wokbook (using **Stat/Regression/Regression**) with this addition:

In the **Options** box: 1.   Enter the x value(s) in the box labeled "Prediction intervals for new observations."

2.   Enter the desired confidence level, such as 95 (for 95% confidence).

If these two steps are followed by using $x = 2.25$, the following display will also be included in addition to the regression equation results.

```
Predicted Values for New Observations

New
Obs     Fit  SE Fit        95% CI              95% PI
  1  2.1609  0.1003  (1.8823, 2.4394)   (1.7202, 2.6015)


Values of Predictors for New Observations

New
Obs  Pizza
  1   2.25
```

The above results show that using the regression equation of $\hat{y} = 0.0346 + 0.945x$, the best predicted value of y (given x = \$2.25) is \$2.1609 (or \$2.16 rounded), and a 95% prediction interval is

$$\$1.72 < y < \$2.60$$

The values shown in the display for "95% C.I." are confidence interval limits for the mean price of subway fare for x = \$2.25.

10-5 Multiple Regression

The textbook discusses multiple regression, and Minitab can provide multiple regression results. Once a collection of sample data has been entered, you can easily experiment with different combinations of columns (variables) to find the combination that is best. Here is the Minitab procedure.

Minitab Procedure for Multiple Regression

1.   Enter the sample data in columns C1, C2, C3, . . . (or open a Minitab worksheet containing the columns of sample data). Also enter the names of the variables in the empty cells located immediately above the first row of sample values.

2.   Click on **Stat**, **Regression**, then **Regression** once again.

3.   Make these entries in the dialog box:
   - Enter the column containing the y values in the box for the "Response" variable.
   - In the box for the "Predictor" variables, enter the columns that you want included for the independent variables (the x variables).
   - Click on the **Options** bar and click on the "**Fit intercept**" box.
   - Click on the **Results** bar and select the second option that includes the regression equation, s, and R-Squared, then click **OK**.

As an example, the table below includes measurements obtained from anesthetized bears. If you enter the sample data included in the table and select column C1 as the response variable and columns C3 and C6 as predictor variables, the Minitab results will be as shown below the table. The results correspond to this multiple regression equation:

$$\hat{y} = -374 + 18.8x_3 + 5.87x_6$$

The results also include the adjusted coefficient of determination (75.9%) as well as a *P*-value (0.012) for overall significance of the multiple regression equation.

Data from Anesthetized Male Bears

Var.	Minitab Column	Name	Sample Data							
y	C1	WEIGHT	80	344	416	348	262	360	332	34
x_2	C2	AGE	19	55	81	115	56	51	68	8
x_3	C3	HEADLEN	11.0	16.5	15.5	17.0	15.0	13.5	16.0	9.0
x_4	C4	HEADWDTH	5.5	9.0	8.0	10.0	7.5	8.0	9.0	4.5
x_5	C5	NECK	16.0	28.0	31.0	31.5	26.5	27.0	29.0	13.0
x_6	C6	LENGTH	53.0	67.5	72.0	72.0	73.5	68.5	73.0	37.0
x_7	C7	CHEST	26	45	54	49	41	49	44	19

Here is the Minitab screen showing the entry of the values in the above table:

→	C1	C2	C3	C4	C5	C6	C7
	WEIGHT	AGE	HEADLEN	HEADWTH	NECK	LENGTH	CHEST
1	80	19	11.0	5.5	16.0	53.0	26
2	344	55	16.5	9.0	28.0	67.5	45
3	416	81	15.5	8.0	31.0	72.0	54
4	348	115	17.0	10.0	31.5	72.0	49
5	262	56	15.0	7.5	26.5	73.5	41
6	360	51	13.5	8.0	27.0	68.5	49
7	332	68	16.0	9.0	29.0	73.0	44
8	34	8	9.0	4.5	13.0	37.0	19

```
The regression equation is
WEIGHT = - 374 + 18.8 HEADLEN + 5.87 LENGTH


Predictor          Coef      SE Coef          T          P
Constant          -374.3       134.1      -2.79      0.038
HEADLEN            18.82       23.15       0.81      0.453
LENGTH             5.875       5.065       1.16      0.299


S = 68.5649     R-Sq = 82.8%     R-Sq(adj) = 75.9%

Analysis of Variance

Source             DF          SS          MS          F          P
Regression          2       113142       56571      12.03      0.012
Residual Error      5        23506        4701
Total               7       136648
```

We can again use Minitab to predict y values for given values of the predictor variables. For example, to predict a bear's weight (y value), given that its head length (HEADLEN) is 14 in. and its LENGTH is 71.0 in., follow the same steps listed earlier in this section, but click on "Options" before clicking on the OK box. In the second dialog box, enter the values of 14.0 and 71.0 in the box identified as "Prediction intervals for new observations," then enter the desired confidence level, such as 95 (for 95% confidence). In addition to the elements in the above Minitab display, the following will also be shown:

```
Predicted Values for New Observations

New Obs    Fit    SE Fit          95.0% CI                95.0% PI
1         306.3     43.9    ( 193.4,   419.2)   (   97.0,   515.6)
```

This portion of the display shows that for a bear with a head length of 14.0 in. and a length of 71.0 in., the predicted weight is 306.3 lb. The prediction interval (for a single weight) is given, and the confidence interval (for the mean weight of all bears with 14.0 in. head length and 71.0 in. length) is also given.

When trying to find the best multiple regression equation, Minitab's **Stepwise Regression** procedure may be helpful. Select **Stat, Regression, Stepwise Regression**.

10-6 Modeling

Mathematical modeling is discussed in the Triola statistics textbooks (except *Essentials of Statistics*). The objective is to find a mathematical function that "fits" or describes real-world data. Among the models discussed in the textbook, we will describe how Minitab can be used for the linear, quadratic, logarithmic, exponential, and power models.

   To illustrate the use of Minitab, consider the sample data in the table below. As in the textbook, we will use the coded year values for x, so that $x = 1, 2, 3, \ldots, 11$. The y values are the populations (in millions) of 5, 10, 17, …, 281. For the following models, we enter 1, 2, 3, . . ., 11 in column C1 and we enter 5, 10, 17, . . . , 281 in column C2. The resulting models are based on the coded values of x (1, 2, 3, . . . , 11) and the population values in millions (5, 10, 17, . . . , 281).

Population (in millions) of the United States

Year	1800	1820	1840	1860	1880	1900	1920	1940	1960	1980	2000
Coded year	1	2	3	4	5	6	7	8	9	10	11
Population	5	10	17	31	50	76	106	132	179	227	281

Linear Model: $y = a + bx$

The linear model can be obtained by using Minitab's correlation and regression module. The procedure is described in Section 10-3 of this manual/workbook. For the data in the above table, enter the coded year values of 1, 2, 3, . . . , 11 in column C1 and enter the population values of 5, 10, 17, . . . , 281 in column C2. The result will be $y = -61.9 + 2.72x$ with $R^2 = 92.5\%$ (or 0.925). The high value of R^2 suggests that the linear model is a reasonably good fit.

Quadratic Model: $y = ax^2 + bx + c$

With the coded values of x in column C1 and the population values (in millions) in column C2, select **Stat**, **Regression**, **Fitted Line Plot**. In the dialog box, enter C2 for the response variable, enter C1 for the predictor variable, and click on the **Quadratic** option. The results show that the function has the form given as $y = 10.01 - 6.003x + 2.767x^2$ with $R^2 = 0.999$. This higher value of R^2 suggests that the quadratic model is a better fit than the linear model.

Logarithmic Model: $y = a + b \ln x$

Minitab does not have a direct procedure for finding a logarithmic model, but there are ways to get it. One approach is to replace the coded x values (1, 2, 3, . . . , 11) with $\ln x$ values (ln 1, ln 2, ln 3, . . . , ln 11) This can be accomplished by using the command editor in the session window; enter the command LET C1 = LOGE(C1). [You could also use **Calc/Calculator** with the expression LOGE(C1).] After replacing the x values with $\ln x$ values, use **Stat**, **Regression**, **Fitted Line Plot** as in Section 10-3. That is, use the same procedure for finding the linear regression equation. The result is $y = -65.89 + 105.1 \ln x$, with $R^2 = 0.696$, suggesting that this model does not fit as well as the linear or quadratic models. Of the three models considered so far, the quadratic model appears to be best.

Exponential Model: $y = ab^x$

The exponential model is tricky, but it can be obtained using Minitab. Enter the values of x in the first column, and enter the values of $\ln y$ in the second column. The values of $\ln y$ can be found by using the command editor in the session window; enter the command LET C2 = LOGE(C2). [You could also use **Calc/Calculator** with the expression LOGE(C2).]

After replacing the y values with $\ln y$ values, use **Stat**, **Regression**, **Regression** to find the linear regression equation as described in Section 10-3. When you get the results from Minitab, the value of the coefficient of determination is correct, but the values of a and b in the exponential model must be computed as follows:

To find the value of a: Evaluate e^{b_0} where b_0 is the y-intercept given by Minitab.

To find the value of b: Evaluate e^{b_1} where b_1 is the slope given by Minitab.

Using the data in the above table, the value of $R^2 = 0.963$ is OK as is, but Minitab's regression equation of $y = 1.6556 + 0.39405x$ must be converted as follows.

$$a = e^{b_0} = e^{1.6556} = 5.2362$$
$$b = e^{b_1} = e^{0.39405} = 1.4830$$

Using these values of a and b, we express the exponential model as $y = 5.2362(1.4830^x)$.

Power Model: $y = ax^b$

Replace the x values with $\ln x$ values and replace the y values with $\ln y$ values. The values of $\ln x$ can be found by using the command editor in the session window; enter the command LET C1 = LOGE(C1). [You could also use **Calc/Calculator** with the expression LOGE(C1).] The values of $\ln y$ can be found the same way. Then use **Stat**, **Regression**, **Regression** to find the linear regression equation as described in Section 10-3 of this workbook. When you get the results from Minitab, the value of the coefficient of determination is correct, the value of b is the same as the slope of the regression line, but the y-intercept must be converted as follows.

To find the value of a: Evaluate e^{b_0} where b_0 is the y-intercept given by Minitab.

Using the data in the above table, we get $R^2 = 0.976$. From Minitab's linear regression equation of $y = 1.2099 + 1.76606x$, we get $b = 1.76606$ and a is computed as follows.

$$a = e^{b_0} = e^{1.2099} = 3.35315$$

Using these values of a and b, we express the power model as $y = 3.35315(x^{1.76606})$.

The rationale underlying the methods for the exponential and power models is based on transformations of equations. In the exponential model of $y = ab^x$, for example, taking natural logarithms of both sides yields $\ln y = \ln a + x (\ln b)$, which is the equation of a straight line. Minitab can be used to find the equation of this straight line that fits the data best; the intercept will be $\ln a$ and the slope will be $\ln b$, but we need the values of a and b, so we solve for them as described above. Similar reasoning is used with the power model.

CHAPTER 10 EXPERIMENTS: Correlation and Regression

10–1. ***Bear Weights and Ages*** Refer to the Bears data set in Appendix B of the textbook. (The Minitab worksheet is BEARS.) Use the values for WEIGHT (x) and the values for AGE (y) to find the following.

  a. Display the scatter diagram of the paired WEIGHT/AGE data. Based on that scatter diagram, does there appear to be a relationship between the weights of bears and their ages? If so, what is it?

  _____

  b. Find the value of the linear correlation coefficient r. _____

  c. Assuming a 0.05 level of significance, what do you conclude about the correlation between weights and ages of bears?

  _____

  d. Find the equation of the regression line. (Use WEIGHT as the x predictor variable, and use AGE as the y response variable.) _____

  e. What is the best predicted age of a bear that weighs 300 lb?_____

10–2. ***Effect of Transforming Data*** The ages used in Experiment 10–1 are in months. Convert them to days by multiplying each age by 30, then repeat Experiment 10–1 and enter the responses here:

  a. Display the scatter diagram of the paired WEIGHT/AGE data. Based on that scatter diagram, does there appear to be a relationship between the weights of bears and their ages? If so, what is it?

  _____

  b. Find the value of the linear correlation coefficient r._____

  c. Assuming a 0.05 level of significance, what do you conclude about the correlation between weights and ages of bears?

  _____

  d. Find the equation of the regression line. (Use WEIGHT as the x predictor variable, and use AGE as the y response variable.) _____

  e. What is the best predicted age of a bear that weighs 300 lb?_____

  f. After comparing the responses obtained in Experiment 10-1 to those obtained here, describe the general effect of changing the scale for one of the variables.

  _____

  _____

10–3. ***Bear Weights and Chest Sizes*** Refer to the Bears data set in Appendix B of the textbook. (The Minitab worksheet is BEARS.) Use the values for CHEST (x) and the values for WEIGHT (y) to find the following.

    a.   Display the scatter diagram of the paired CHEST/WEIGHT data. Based on that scatter diagram, does there appear to be a relationship between the chest sizes of bears and their weights? If so, what is it?

    b.   Find the value of the linear correlation coefficient r. _____

    c.   Assuming a 0.05 level of significance, what do you conclude about the correlation between chest sizes and weights of bears?

    d.   Find the equation of the regression line. (Use CHEST as the x predictor variable, and use WEIGHT as the y response variable.) _____

    e.   What is the best predicted weight of a bear with a chest size of 36.0 in?_____

    f.   When trying to obtain measurements from an anesthetized bear, what is a practical advantage of being able to predict the bear's weight by using its chest size?

10–4. ***Effect of No Variation for a Variable*** Use the following paired data and obtain the indicated results.

x | 1 2 3 4 5 7 7 9
y | 5 5 5 5 5 5 5 5

    a.   Print a scatter diagram of the paired x and y data. Based on the result, does there appear to be a relationship between x and y? If so, what is it?

    b.   What happens when you try to find the value of r? Why?

    c.   What do you conclude about the correlation between x and y? What is the equation of the regression line?

10–5. ***CPI and Pizza*** The paired values of the Consumer Price Index (CPI) and the cost of a slice of pizza are listed below. Is there a linear correlation between the CPI and the cost of a slice of pizza?

CPI	30.2	48.3	112.3	162.2	191.9	197.8
Cost of Pizza	0.15	0.35	1.00	1.25	1.75	2.00

Is there a correlation between the CPI and the cost of a slice of pizza? Explain.

_____

What is the equation of the regression line? _____

Find the best predicted value for the cost of a slice of pizza given a CPI of 200.

_____

10–6. ***Supermodel Heights and Weights*** Listed below are heights (in inches) and weights (in pounds) for supermodels Niki Taylor, Nadia Avermann, Claudia Schiffer, Elle MacPherson, Christy Turlington, Bridget Hall, Kate Moss, Valerie Mazza, and Kristy Hume.

Height (in.)	71	70.5	71	72	70	70	66.5	70	71
Weight (lb.)	125	119	128	128	119	127	105	123	115

Is there a correlation between height and weight? If there is a correlation, does it mean that there is a correlation between height and weight of all adult women?

_____

What is the equation of the regression line? _____

Find the best predicted weight of a supermodel who is 69 in. tall. _____

10–7. ***Movie Data***   Refer to the data from movies in the table below, where all amounts are in millions of dollars. Let the movie gross amount be the dependent *y* variable.

Budget	62	90	50	35	200	100	90
Gross	65	64	48	57	601	146	47

Is there a correlation between the budget amount and the gross amount? Explain.

_____

What is the equation of the regression line? _____

Find the best predicted gross amount for a movie with a budget of $150 million._____

10–8. *Crickets and Temperature* The numbers of chirps in one minute were recorded for different crickets, and the corresponding temperatures were also recorded. The results are given in the table below.

Chirps in one minute	882	1188	1104	864	1200	1032	960	900
Temperature (°F)	69.7	93.3	84.3	76.3	88.6	82.6	71.6	79.6

Is there a correlation between the number of chirps and the temperature? Explain.

_____

What is the equation of the regression line? _____

Find the best predicted temperature when a cricket chirps 1000 times in one minute.___

10–9. *Blood Pressure* Refer to the systolic and diastolic blood pressure measurements from randomly selected subjects. Let the dependent variable y represent diastolic blood pressure.

Systolic	138	130	135	140	120	125	120	130	130	144	143	140	130	150
Diastolic	82	91	100	100	80	90	80	80	80	98	105	85	70	100

Is there a correlation between systolic and diastolic blood pressure? Explain.

_____

What is the equation of the regression line? _____

Find the best predicted measurement of diastolic blood pressure for a person with a systolic blood pressure of 123. _____

10–10. *Garbage Data for Predicting Household Size* The "Garbage" data set in Appendix B of the textbook consists of data from the Garbage Project at the University of Arizona. (The Minitab worksheet is GARBAGE.) Use household size (HHSIZE) as the response y variable. For each predictor x variable given below, find the value of the linear correlation coefficient, the equation of the regression line, and the value of the coefficient of determination r^2. Enter the results in the spaces below.

(*continued*)

	_r_	Equation of regression line	r^2
Metal	___	_____	___
Paper	___	_____	___
Plastic	___	_____	___
Glass	___	_____	___
Food	___	_____	___
Yard	___	_____	___
Text	___	_____	___
Other	___	_____	___
Total	___	_____	___

Based on the above results, which single independent variable appears to be the best predictor of household size? Why?

_____

_____

10–11. Use the same data set described in Experiment 10-10. Let household size (HHSIZE) be the dependent _y_ variable and use the given predictor _x_ variables to fill in the results below.

	Multiple regression eq.	R^2	Adj. R^2
Metal and Paper	_____	___	___
Plastic and Food	_____	___	___
Metal, Paper, Glass	_____	___	___
Metal, Paper, Plastic, Glass	_____	___	___

Based on the above results, which of the multiple regression equations appears to best fit the data? Why?

_____

_____

10-12. *Appendix B Data Set: Predicting Nicotine in Cigarettes* Refer to Data Set 4 in Appendix B and use the tar, nicotine, and CO amounts for the cigarettes that are 100 mm long, filtered, non-menthol, and non-light (the last set of measurements). Find the best regression equation for predicting the amount of nicotine in a cigarette. Is the best regression equation a good regression equation for predicting the nicotine content?

_____

_____

10-13. *Manatee Deaths from Boats* Listed below are the numbers of Florida manatee deaths related to encounters with watercraft (based on data from *The New York Times*). The data are listed in order, beginning with the year 1980 and ending with the year 2000.

16 24 20 15 34 33 33 39 43 50 47 53 38 35 49 42 60 54 67 82 78

Use the given data to find equations and coefficients of determination for the indicated models.

	Equation	R^2
Linear	_____	_____
Quadratic	_____	_____
Logarithmic	_____	_____
Exponential	_____	_____
Power	_____	_____

Based on the above results, which model appears to best fit the data? Why?

_____

What is the best predicted value for 2001? In 2001, there were 82 watercraft–related manatee deaths. How does the predicted value compare to the actual value?

_____

11

Goodness-of-Fit and Contingency Tables

11-1 Goodness-of-Fit

We use a *goodness-of-fit* test with sample data consisting of observed frequency counts arranged in a single row or column (called a one-way frequency table). We test for the claim that the observed frequency counts agree with some claimed distribution, so that there is a *good fit* of the observed data with the claimed distribution.

Minitab Procedure for Goodness-of-Fit

1. First collect the observed values O and enter them in column C1.

2. If the expected frequencies are not all the same, enter the expected *proportions* in column C2.

3. Select **Stat**, **Tables**, and **Chi-Square Goodness-of-Fit Test (One Variable)**..

4. Enter C1 in the "Observed counts" box.

5. Select "Equal proportions" if the expected frequencies are all the same. Otherwise, select "Specific proportions" and enter C2 in the box.

6. Click **OK.**

The textbook includes an example with the frequency counts of the 80 last digits of the weights of randomly selected adults. Those frequency counts are listed in the table below. We want to test the claim that the observed digits are from a population of weights in which the last digits do not occur with the same frequency.

Last Digits of Weights

Last Digit	Frequency
0	7
1	14
2	6
3	10
4	8
5	4
6	5
7	6
8	12
9	8

After entering the observed frequencies column C1, select **Stat**, **Tables,** select **Chi-Square Goodness-of-Fit Test (One Variable)** and use the dialog box as shown below.

Chi-Square Goodness-of-Fit Test

C1

- ⦿ Observed counts: C1
-    Category names (optional) :
- ◯ Categorical data:

Test
- ⦿ Equal proportions
- ◯ Specific proportions
- ◯ Proportions specified by historical counts:
  - Input column ▾

Select   Graphs...   Results...

Help   OK   Cancel

The Minitab results are shown below.

Chi-Square Goodness-of-Fit Test for Observed Counts in Variable: C1

Category	Observed	Test Proportion	Expected	Contribution to Chi-Sq
1	7	0.1	8	0.125
2	14	0.1	8	4.500
3	6	0.1	8	0.500
4	10	0.1	8	0.500
5	8	0.1	8	0.000
6	4	0.1	8	2.000
7	5	0.1	8	1.125
8	6	0.1	8	0.500
9	12	0.1	8	2.000
10	8	0.1	8	0.000

N	DF	Chi-Sq	**P-Value**
80	9	11.25	**0.259**

The *P*-value of 0.259 indicates that we fail to reject the null hypothesis that the digits occur with the same frequency. There is not sufficient evidence to support the claim that the digits do not occur with the same frequency.

11-2 Contingency Tables

A **contingency table** (or **two-way frequency table**) is a table in which frequencies correspond to two variables. One variable is used to categorize rows, and a second variable is used to categorize columns. Let's consider the data in the contingency table shown below. The row variable identifies whether the subjects became infected, and the column variable identifies the treatment group (placebo, 20% extract group, or 60% extract group).

Results from Experiment with Echinacea

	Placebo	Echinacea: 20% extract	Echinacea: 60% extract
		Treatment Group	
Infected	88	48	42
Not infected	15	4	10

Minitab can analyze data from a contingency table. To use Minitab with a contingency table, follow this procedure:

Minitab Procedure for Contingency Tables

1. Enter the individual columns of data in Minitab columns C1, C2, C3, . . . , .
    For the data in the above table, enter 88 and 15 in column C1, then enter 48 and 4 in column C2, and enter 42 and 10 in column C3. From the Minitab screen below, we can see that the table is entered as it appears.

↓	C1	C2	C3
1	88	48	42
2	15	4	10

2. Select **Stat, Tables, Chisquare Tes (Two-Way Table in Worksheet)**.

3. Proceed to enter the names of the columns containing the observed frequencies. For the above data, enter C1 C2 C3. Click **OK**.

Shown below are the Minitab results from the data in the above table. The *P*-value has been highlighted in bold, and it is 0.232. If using a 0.05 significance level, we fail to reject the null hypothesis of independence between the row and column variables. It appears that getting an infection is independent of the treatment group. This suggests that echinacea is not an effective treatment for colds. The display also includes the test statistic of 2.925 as well as the expected value for each cell and the contribution of each cell to the total value of the chi-square test statistic.

Chi-Square Test: C1, C2, C3

```
Expected counts are printed below observed counts
Chi-Square contributions are printed below expected counts

           C1       C2       C3    Total
    1      88       48       42     178
         88.57    44.71    44.71
         0.004    0.241    0.165

    2      15        4       10      29
         14.43     7.29     7.29
         0.023    1.481    1.012

Total    103       52       52     207

Chi-Sq = 2.925, DF = 2, P-Value = 0.232
```

11-3 Fisher's Exact Test

Fisher's exact test can be used for two-way tables. This test uses an *exact* distribution instead of an approximating chi-square distribution. It is particularly helpful when the approximating chi-square distribution cannot be used because of expected cell frequencies that are less than 5. Consider the sample data in the table below, with expected frequencies shown in parentheses. Note that the first cell has an expected frequency of 3, which is less than 5, so the chi-square distribution should not be used.

Helmets and Facial Injuries in Bicycle Accidents

	Helmet Worn	No Helmet
Facial injuries received	2 (3)	13 (12)
All injuries nonfacial	6 (5)	19 (20)

Minitab procedure for Fisher's Exact Test

1.   Enter the frequencies in column C1 with the corresponding row numbers in column C2 and the corresponding column numbers in column C3. Here are the Minitab entries for the above table. (The column labels of Frequency, Row, and Column were manually entered.) The first frequency of 2 is located in row 1 and column 1, as indicated by the entries in Minitab columns C2 and C3. The second frequency of 13 is in row 1 and column 2, and so on.

↓	C1	C2	C3
	Frequency	Row	Column
1	2	1	1
2	13	1	2
3	6	2	1
4	19	2	2

2.   Select **Stat**, **Tables**, then **Cross Tabulation and Chi-Square**.

3.   Enter C2 in the "For rows" box, enter C3 in the "For columns" box, and enter C1 in the "Frequencies are in" box.

4.   Click on the **Other Stats** bar and select "Fisher's exact test for 2×2 tables."

5.   Click **OK** twice to get the resulting P-value. Here is the display that results from the above table: **Fisher's exact test: P-Value = 0.685661.** Because the P-value is large, we fail to reject the null hypothesis that wearing a helmet and receiving facial injuries are independent.

CHAPTER 11:   Goodness-of-Fit and Contingency Tables

11-1.   ***Loaded Die*** The author drilled a hole in a die and filled it with a lead weight, then proceeded to roll it 200 times. Here are the observed frequencies for the outcomes of 1, 2, 3, 4, 5, and 6 respectively: 27, 31, 42, 40, 28, 32. Use a 0.05 significance level to test the claim that the outcomes are not equally likely.

Test statistic:_____   *P*-value:_____

Conclusion:_____

_____

Does it appear that the loaded die behaves differently than a fair die?

_____

11–2.   ***Flat Tire and Missed Class*** A classic tale involves four car-pooling students who missed a test and gave as an excuse a flat tire. On the makeup test, the instructor asked the students to identify the particular tire that went flat. If they really didn't have a flat tire, would they be able to identify the same tire? The author asked 41 other students to identify the tire they would select. The results are listed in the following table (except for one student who selected the spare). Use a 0.05 significance level to test the author's claim that the results fit a uniform distribution.

Tire	Left front	Right front	Left rear	Right rear
Number selected	11	15	8	6

Test statistic:_____   *P*-value:_____

Conclusion:_____

_____

What does the result suggest about the ability of the four students to select the same tire when they really didn't have a flat?

_____

11-3. ***Occupational Injuries*** Randomly selected non-fatal occupational injuries and illnesses are categorized according to the day of the week that they first occurred, and the results are listed below (based on data from the Bureau of Labor Statistics). Use a 0.05 significance level to test the claim that such injuries and illnesses occur with equal frequency on the different days of the week.

Day	Mon	Tues	Wed	Thurs	Fri
Number	23	23	21	21	19

Test statistic:_____   *P*-value:_____

Conclusion:_____

_____

11-4. ***Measuring Pulse Rates*** According to one procedure used for analyzing data, when certain quantities are measured, the last digits tend to be uniformly distributed, but if they are estimated or reported, the last digits tend to have disproportionately more 0s or 5s. Refer to Data Set 1 in Appendix B of the textbook and use the last digits of the pulse rates of the 80 men and women. Those pulse rates were obtained as part of the National Health and Examination Survey. Test the claim that the last digits of 0, 1, 2, 3, . . . , 9 occur with the same frequency.

Test statistic:_____   *P*-value:_____

Conclusion:_____

_____

What can be inferred about the procedure used to obtain the pulse rates?

_____

11-5. ***Genetics Experiment*** Based on the genotypes of parents, offspring are expected to have genotypes distributed in such a way that 25% have genotypes denoted by AA, 50% have genotypes denoted by Aa, and 25 % have genotypes denoted by aa. When 145 offspring are obtained, it is found that 20 of them have AA genotypes, 90 have Aa genotypes and 35 have aa genotypes. Test the claim that the observed genotype offspring frequencies fit the expected distribution of 25% for AA, 50% for Aa, and 25% for aa. Use a significance level of 0.05.

Test statistic:_____   *P*-value:_____

Conclusion:_____

_____

11-6. ***Testing a Normal Distribution***   In this experiment we will use Minitab's ability to generate normally distributed random numbers. We will then test the sample data to determine if they actually do fit a normal distribution.

a. Generate 1000 random numbers from a normal distribution with a mean of 100 and a standard deviation of 15. (IQ scores have these parameters.) Select **Calc**, then **Random Data**, then **Normal**.

b. Use **Data/Sort** (or **Manip/Sort** on Minitab Release 13 and earlier) to arrange the data in order.

c. Examine the sorted list and determine the frequency for each of the categories listed below. Enter those frequencies in the spaces provided. (The expected frequencies are found by assuming that the data are normally distributed.)

	Observed Frequency	Expected Frequency
Below 55:	_____	1
55-70:	_____	22
70-85:	_____	136
85-100:	_____	341
100-115:	_____	341
115-130:	_____	136
130-145:	_____	22
Above 145:	_____	1

d. Use Minitab to test the claim that the randomly generated numbers actually do fit a normal distribution with mean 100 and standard deviation 15.

Test statistic:_____   *P*-value:_____

Conclusion:_____

_____

11-7. ***Instant Replay in Tennis*** The table below summarizes challenges made by tennis players in the first U. S. Open that used the Hawk-Eye electronic instant replay system. Use a 0.05 significance level to test the claim that success in challenges is independent of the gender of the player. Does either gender appear to be more successful?

	Was the challenge to the call successful?	
	Yes	No
Men	201	288
Women	126	224

Test statistic:_____   *P*-value:_____

Conclusion:_____

_____

11-8. ***Accuracy of Polygraph Tests*** The data in the accompanying table summarize results from tests of the accuracy of polygraphs (based on data from the Office of Technology Assessment). Use a 0.05 significance level to test the claim that whether the subject lies is independent of the polygraph indication.

	Polygraph Indicated Truth	Polygraph Indicated Lie
Subject actually told the truth	65	15
Subject actually told a lie	3	17

Test statistic:_____   *P*-value:_____

Conclusion:_____

_____

What do the results suggest about the effectiveness of polygraphs?

_____

_____

11–9. ***Global Warming Survey*** A Pew Research poll was conducted to investigate opinions about global warming. The respondents who answered yes when asked if there is solid evidence that the earth is getting warmer were then asked to select a cause of global warming. The results are given the table below. Use a 0.05 significance level to test the claim that the sex of the respondent is independent of the choice for the cause of global warming. Do men and women appear to agree, or is there a substantial difference?

	Human activity	Natural patterns	Don't know or refused to answer
Male	314	146	44
Female	308	162	46

Test statistic:_____   *P*-value:_____

Conclusion:_____

_____

11–10. ***Occupational Hazards*** Use the data in the table to test the claim that occupation is independent of whether the cause of death was homicide. The table is based on data from the U.S. Department of Labor, Bureau of Labor Statistics.

	Police	Cashiers	Taxi Drivers	Guards
Homicide	82	107	70	59
Cause of Death Other than Homicide	92	9	29	42

Test statistic:_____   *P*-value:_____

Conclusion:_____

_____

Does any particular occupation appear to be most prone to homicides? If so, which one?

_____

How are the results affected if the order of the rows is switched?

_____

How are the results affected by the presence of an outlier? If we change the first entry from 82 to 8200, are the results dramatically affected?

_____

11-11. ***Fisher's Exact Test***   Refer to Experiment 11-7 in this manual/workbook. Repeat that experiment by using Fisher's exact test instead of using the approximating chi-square distribution. Enter the results below.

P-value obtained by using the approximating chi-square distribution: _____

P-value obtained by using Fisher's exact test:

_____

Does the use of the Fisher's exact test have much of an effect on the P-value?

_____

_____

11-12. ***Fisher's Exact Test***   Refer to Experiment 11-8 in this manual/workbook. Repeat that experiment by using Fisher's exact test instead of using the approximating chi-square distribution. Enter the results below.

P-value obtained by using the approximating chi-square distribution: _____

P-value obtained by using Fisher's exact test:

_____

Does the use of the Fisher's exact test have much of an effect on the P-value?

_____

_____

12

Analysis
of
Variance

12-1 One-Way Analysis of Variance

One–way analysis of variance is used to test the claim that three or more populations have the same mean. When the Triola textbook discusses one-way analysis of variance, it is noted that the term "one-way" is used because the sample data are separated into groups according to one characteristic or "factor". In the table below, the factor is the size of the car (small, medium, or large).

Chest Deceleration Measurements (in g's) from Car Crash Tests

Small Cars   44 43 44 54 38 43 42 45 44 50

Medium Cars   41 49 43 41 47 42 37 43 44 34

Large Cars   32 37 38 45 37 33 38 45 43 42

Because the calculations are very complicated, the Triola textbook emphasizes the interpretation of results obtained by using software, so Minitab is suitable for this topic. We should understand that a small P-value (such as 0.05 or less) leads to rejection of the null hypothesis of equal means. ["If the P (value) is low, the null must go."] With a large P-value (such as greater than 0.05), fail to reject the null hypothesis of equal means.

The following procedure describes how Minitab can be used with such collections of sample data to test the claim that the different samples come from populations with the same mean. For the data in the above table, the claim of equal means leads to these hypotheses:

H_0: $\mu_1 = \mu_2 = \mu_3$

H_1: At least one of the population means is different from the others.

Minitab Procedure for One-Way Analysis of Variance

1.   Enter the original sample values in columns (or open a Minitab worksheet that contains the columns of data).

2.   Select **Stat** from the main menu.

3.   Select the subdirectory item of **ANOVA**.

4.   Select the option of **Oneway (Unstacked)**. The term "unstacked" means that the data are not all stacked in one single column; they are listed in their individual and separate columns.

5.   In the dialog box, enter the column names (C1 C2 C3) in the box labeled as "Responses (in separate columns)." Click **OK**.

If you use the above steps with the chest deceleration measurements from the above table, the Minitab results will appear as shown below.

One-way ANOVA: Small, Medium, Large

```
Source   DF     SS    MS     F     P
Factor    2   162.9  81.4  4.09  0.028
Error    27   537.0  19.9
Total    29   699.9

S = 4.460    R-Sq = 23.27%    R-Sq(adj) = 17.59%

                         Individual 95% CIs For Mean Based on Pooled StDev
Level    N    Mean   StDev    +---------+---------+---------+---------
Small   10   44.700  4.398                          (---------*---------)
Medium  10   42.100  4.358                 (--------*---------)
Large   10   39.000  4.619      (---------*---------)
                              +---------+---------+---------+---------
                             36.0      39.0      42.0      45.0

Pooled StDev = 4.460
```

The *P*-value of 0.028 indicates that there is sufficient sample evidence to warrant rejection of the null hypothesis that $\mu_1 = \mu_2 = \mu_3$. The test statistic of $F = 4.09$ is also provided, as are the values of the SS and MS components. The lower portion of the Minitab display also includes the individual sample means, standard deviations, and graphs of the 95% confidence interval estimates of each of the three population means. These confidence intervals are computed by the same methods used in Chapter 7, except that a pooled standard deviation is used instead of the individual sample standard deviations. (The last entry in the Minitab display shows that the pooled standard deviation is 4.460.)

Note that the Minitab display uses the term *factor* instead of *treatment*, so that the value of SS(treatment) is 162.9, and MS(treatment) = 81.4

12-2 Two-Way Analysis of Variance

Two-way analysis of variance involves *two* factors, such as type of car (foreign or domestic) and size (small, medium, large) in the table below. The two–way analysis of variance procedure requires that we test for (1) an interaction effect between the two factors; (2) an effect from the row factor; (3) an effect from the column factor.

Chest Deceleration Measurements (in g's) from Car Crash Tests

	Size of Car		
	Small	Medium	Large
Foreign	44	41	32
	54	49	45
	43	47	42
Domestic	43	43	37
	44	37	38
	42	34	33

Entering the Sample Data The first obstacle to overcome is to understand a somewhat awkward method of entering sample data. When entering the values in the above table, we must somehow keep track of the location of each data value, so here is the procedure to be used:

1. Enter *all* of the sample data in column C1.

2. Enter the corresponding row numbers in column C2.

3. Enter the corresponding column numbers in column C3.

This particular format can be confusing at first, so stop and try to recognize the pattern. Because the first value of 44 from the above table is in the first row and first column, its row identification is 1 and its column identification is 1. The entry of 54 is also in the first row and first column, so its row identification number is 1 and its column identification number is 2. (The sample values of 44, 54, and 43 are all considered to be in the first row (foreign) and the first column (small). The value of 41 is in the first row and second column.

Also, it's wise to name column C1 as "CHEST," name column C2 as "TYPE" (for type of car: foreign or domestic), and name column C3 as "SIZE" so that the columns have meaningful names. Here is an illustration of the pattern used to enter the sample data in the above table.

See the following page where the above table is entered along with the corresponding Minitab worksheet window.

Chest Deceleration Measurements (in g's) from Car Crash Tests

	Size of Car		
	Small	Medium	Large
Foreign	44	41	32
	54	49	45
	43	47	42
Domestic	43	43	37
	44	37	38
	42	34	33

↓

↓	C1 CHEST	C2 TYPE	C3 SIZE
1	44	1	1
2	54	1	1
3	43	1	1
4	41	1	2
5	49	1	2
6	47	1	2
7	32	1	3
8	45	1	3
9	42	1	3
10	43	2	1
11	44	2	1
12	42	2	1
13	43	2	2
14	37	2	2
15	34	2	2
16	37	2	3
17	38	2	3
18	33	2	3

The procedure for obtaining a Minitab display for two-way analysis of variance is as follows.

Minitab Procedure for Two-Way Analysis of Variance

1. Enter all of the sample values in column C1 (as shown above).

2. Enter the corresponding row numbers in column C2 (as shown above).

3. Enter the corresponding column numbers in column C3 (as shown above).

4. Select **Stat**, then **ANOVA**, then **Twoway**.

5. You should now see a dialog box like the one shown below.

   Make these entries in the dialog box:

   - Enter C1 (or its column label) in the box labeled Response.
   - Enter C2 (or its column label) in the box labeled Row Factor.
   - Enter C3 (or its column label) in the box labeled Column Factor.
   - Click **OK**.

Using the data in the preceding table will result in the Minitab display shown below.

Two-way ANOVA: CHEST versus TYPE, SIZE

```
Source          DF        SS        MS       F       P
TYPE             1   117.556   117.556    5.44   0.038
SIZE             2   154.778    77.389    3.58   0.060
Interaction      2    14.778     7.389    0.34   0.717
Error           12   259.333    21.611
Total           17   546.444

S = 4.649    R-Sq = 52.54%    R-Sq(adj) = 32.77%
```

The Triola textbooks (excluding *Essentials of Statistics*) describe the interpretation of the preceding Minitab display.

1. Test for Interaction
We begin by testing the null hypothesis that there is no *interaction* between the two factors of SITE and TREATMENT. Using the above Minitab results, we calculate the following test statistic.

$$F = \frac{MS(interaction)}{MS(error)} = \frac{7.389}{21.611} = 0.34$$

Minitab shows the corresponding *P*-value of 0.717, so we fail to reject the null hypothesis of no interaction between the two factors of TYPE and SIZE. There does not appear to be an effect from an interaction between type (foreign, domestic) and size (small, medium, large).

2. Test for Effect from the Row Factor
Our two-way analysis of variance procedure outlined in the textbook indicates that we should now proceed to test these two null hypotheses: (1) There are no effects from the row factor (TYPE); (2) There are no effects from the column factor (SIZE). For the test of an effect from the row factor (TYPE), we have

$$F = \frac{MS(TYPE)}{MS(error)} = \frac{117.556}{21.611} = 5.44$$

The Minitab *P*-value is 0.038, so we reject the claim that there is no effect from type of car. That is, the type of car (foreign, domestic) does appear to have an effect on the chest deceleration measurement.

3. Test for Effect from the Column Factor

For the test of an effect from SIZE, we have

$$F = \frac{MS(SIZE)}{MS(error)} = \frac{77.389}{21.611} = 3.58$$

The Minitab P–value is displayed as 0.060, so size does not appear to have an effect on chest deceleration measurement.

Special Case: One Observation Per Cell

The Triola textbook (excluding *Essentials of Statistics*) includes a subsection describing the special case in which there is only one sample value in each cell. Here's how we proceed when there is one observation per cell: *If it seems reasonable to assume (based on knowledge about the circumstances) that there is no interaction between the two factors, make that assumption and then proceed as before to test the following two hypotheses separately:*

H_0: There are no effects from the row factor.

H_0: There are no effects from the column factor.

To use Minitab, simply apply the same procedure described earlier. The Minitab results will not include values for an interaction, but the other necessary values are provided.

CHAPTER 12 EXPERIMENTS: Analysis of Variance

12-1. ***Head Injury in a Car Crash*** Listed below are head injury data from car crash test
dummies. These measurements are in hic, which denotes a standard head injury criterion.
Use a 0.05 significance level to test the null hypothesis that the different car categories
have the same mean. Do these data suggest that larger cars are safer?

Small Cars   290 406 371 544 374 501 376 499 479 475

Medium Cars 245 502 474 505 393 264 368 510 296 349

Large Cars   342 216 335 698 216 169 608 432 510 332

SS(treatment): _____   MS(treatment): _____   Test statistic F: _____

SS(error):   _____   MS(error):   _____ P-value: _____

SS(total):   _____

Conclusion:_____

_____

12-2. ***Femur Injury in a Car*** Crash Listed below are measured loads (in lb) on the left femur
of car crash test dummies. Use a 0.05 significance level to test the null hypothesis that the
different car categories have the same mean. Do these data suggest that larger cars are
safer?

Small Cars   548 782 1188 707 324 320 634 501 274 437

Medium Cars 194 280 1076 411 617 133 719 656 874 445

Large Cars   215 937 953 1636 937 472 882 562 656 433

SS(treatment): _____   MS(treatment): _____   Test statistic F: _____

SS(error):   _____   MS(error):   _____ P-value: _____

SS(total):   _____

Conclusion:_____

_____

12–3. *Archeology: Skull Breadths from Different Epochs* The values in the table are measured maximum breadths of male Egyptian skulls from different epochs (based on data from Ancient Races of the Thebaid, by Thomson and Randall-Maciver). Changes in head shape over time suggest that interbreeding occurred with immigrant populations. Use a 0.05 significance level to test the claim that the different epochs do not all have the same mean.

4000 B.C.	1850 B.C.	150 A.D.
131	129	128
138	134	138
125	136	136
129	137	139
132	137	141
135	129	142
132	136	137
134	138	145
138	134	137

SS(treatment): _____   MS(treatment): _____   Test statistic F: _____

SS(error): _____   MS(error): _____ P-value: _____

SS(total): _____

Conclusion:_____

_____

12–4. *Mean Weights of M&Ms* Refer to the M&M data set in Appendix B from the textbook. (The Minitab worksheet name is M&M.) At the 0.05 significance level, test the claim that the mean weight of M&Ms is the same for each of the six different color populations.

SS(treatment): _____   MS(treatment): _____   Test statistic F: _____

SS(error): _____   MS(error): _____ P-value: _____

SS(total): _____

Conclusion:_____

If it is the intent of Mars, Inc., to make the candies so that the different color populations have the same mean weight, do these results suggest that the company has a problem requiring corrective action?

_____

12–5.   ***Nicotine in Cigarettes*** Refer to Data Set 4 in Appendix B and use the amounts of nicotine (mg per cigarette) in the king size cigarettes, the 100 mm menthol cigarettes, and the 100 mm non-menthol cigarettes. The king size cigarettes are non-filtered, non-menthol, and non-light. The 100 mm menthol cigarettes are filtered and non-light. The 100 mm non-menthol cigarettes are filtered and non-light. Use a 0.05 significance level to test the claim that the three categories of cigarettes yield the same mean amount of nicotine. Given that only the king size cigarettes are not filtered, do the filters appear to make a difference?

SS(treatment): _____   MS(treatment): _____   Test statistic F: _____

SS(error): _____   MS(error): _____ P-value: _____

SS(total): _____

Conclusion: _____

12–6.   ***Tar in Cigarettes*** Refer to Data Set 4 in Appendix B and use the amounts of tar (mg per cigarette) in the three categories of cigarettes described in Exercise 15. Use a 0.05 significance level to test the claim that the three categories of cigarettes yield the same mean amount of tar. Given that only the king size cigarettes are not filtered, do the filters appear to make a difference?

SS(treatment): _____   MS(treatment): _____   Test statistic F: _____

SS(error): _____   MS(error): _____ P-value: _____

SS(total): _____

Conclusion: _____

12-7.   ***Simulations*** Use Minitab to randomly generate three different samples of 500 values each. (Select **Calc, Random Data, Normal**.) For the first two samples, use a normal distribution with a mean of 100 and a standard deviation of 15. For the third sample, use a normal distribution with a mean of 101 and a standard deviation of 15. We know that the three populations have different means, but do the methods of analysis of variance allow you to conclude that the means are different? Explain.

12–8. *Pulse Rates* Use the following pulse rates from Data Set 1 in Appendix B.

	Age		
	Under 20	20–40	Over 40
Male	96 64 68 60	64 88 72 64	68 72 60 88
Female	76 64 76 68	72 88 72 68	60 68 72 64

Are pulse rates affected by an interaction between gender and age? Explain.

_____

Are pulse rates affected by gender? Explain.

_____

Are pulse rates affected by age? Explain.

_____

_____

12-9. *Cholesterol Levels* The following table lists measured cholesterol levels from Data Set 1 in Appendix B. Are cholesterol levels affected by an interaction between sex and age? Are cholesterol levels affected by sex? Are cholesterol levels affected by age?

	Age		
	Under 30	30–50	Over 50
Male	265 303 1252 230 957	702 277 176 416 120	75 189 288 578 31
Female	325 112 62 301 223	146 173 149 462 94	146 173 149 462 94

Are cholesterol levels affected by an interaction between gender and age? Explain.

_____

Are cholesterol levels affected by gender? Explain.

_____

Are cholesterol levels affected by age? Explain.

_____

_____

13

Nonparametric Statistics

13-1 Ranking Data

This chapter describes how Minitab can be used for the nonparametric methods presented in Chapter 13 of *Elementary Statistics,* 11th edition. The sections of this chapter correspond to those in the textbook. The textbook introduces some basic principles of nonparametric methods, but it also describes a procedure for converting data into their corresponding *ranks*. Here is the Minitab procedure for using data in column C1 to create a column C2 that consists of the corresponding ranks.

Converting Data to Ranks

1. Enter the data values in column C1.

2. Select **Data** from the main menu.

3. Select **Rank** from the subdirectory.

4. In the dialog box, make these entries:

   • For the box labeled "Rank data in," enter C1.

   • For the box labeled "Store ranks in," enter C2.

   • Click **OK**.

As an example, consider the values of 5, 3, 40, 10, 12, and 12. They are converted to the ranks of 2, 1, 6, 3, 4.5, and 4.5. Enter the values of 5, 3, 40, 10, 12, and 12 in column C1, follow the above steps, and column C2 will contain the ranks of 2, 1, 6, 3, 4.5, and 4.5 as shown below.

↓	C1	C2
1	5	2.0
2	3	1.0
3	40	6.0
4	10	3.0
5	12	4.5
6	12	4.5

Ties: Note that Minitab handles ties as described in the Triola textbook. See the above display that includes ties in the ranks that correspond to the values of 12 and 12. Ties occur with the corresponding ranks of 4 and 5, so Minitab assigns a rank of 4.5 to each of those values.

13-2 Sign Test

The Triola textbook (excluding *Essentials of Statistics*) includes the following definition.

Definition

The **sign test** is a nonparametric (distribution-free) test that uses plus and minus signs to test different claims, including:
1. Claims involving matched pairs of sample data
2. Claims involving nominal data
3. Claims about the median of a single population

Minitab makes it possible to work with all three of the above cases. Let's consider the matched data in the table below. (The data are matched, because each pair of values is from *adjacent* plots of land.)

Weight (kg) Measurements of Students in Their Freshman Year

September weight	67	53	64	74	67	70	55	74	62	57
April weight	66	52	68	77	67	71	60	82	65	58
Sign of difference	+	+	−	−	0	−	−	−	−	−

Given such paired data, we can use Minitab to apply the sign test to test the claim of no difference.

Minitab Procedure for Sign Test with Matched Data

1. Enter the values of the first variable in column C1.

2. Enter the values of the second variable in column C2.

3. Create column C3 equal to the values of C1 − C2. Accomplish this using either of the following two approaches.

   Session window: Click on the session window at the top portion of the screen, then click on **Editor**, then click on **Enable Commands**. Enter the command:
   LET C3 = C1 − C2.

   Calculator: Click on **Calc**, then **Calculator**. In Minitab's calculator, enter C3 for the "Store result in variable" box, and enter the expression C1 − C2 in the "expression" box. Click **OK** when done.

4. Select **Stat** from the main menu.

5. Select **Nonparametrics** from the subdirectory.

6. Select the option of **1-Sample Sign**.

7. Make these entries in the resulting dialog box:

  - For the box labeled Variables, enter C3 (which contains the differences obtained when column C2 is subtracted from column C1).

  - Click on the small circle for **Test median** and, in the adjacent box, enter 0 for a test of the claim that the median is equal to zero.

  - In the box labeled Alternative, select **not equal,** so that the alternative hypothesis is that the statement that the median is not equal to zero.

  - Click **Ok**.

Here are the Minitab results for the data in the above table:

```
Sign test of median =  0.00000 versus not = 0.00000

     N  Below  Equal  Above       P  Median
C3  10      7      1      2  0.1797  -2.000
```

The above results show that there are 10 pairs of data, there are 7 differences that are below zero (negative sign), one of the differences is equal to zero, and one of the differences are above zero (positive sign). The P-value is 0.1797. In the textbook, we would use a decision criterion that involves a comparison of the test statistic $x = 1$ (the less frequent sign) and a critical value found in Table A-7. Because Minitab displays the P-value, we can use the P-value approach to hypothesis testing. Because the P-value of 0.1797 is greater than the significance level of $\alpha = 0.05$, there is not sufficient evidence to warrant rejection of the claim that the median of the differences is equal to 0. There does not appear to be significant difference between the September weights and the April weights.

Sign test with nominal data: In some cases, the data for the sign test do not consist of values of the type given in the table above. For example, suppose that a test of a gender-selection method results in 726 babies, including 668 girls. Because the data are at the nominal level of measurement, we could enter them as −1's (for girls) and 1's (for boys). *Shortcut*: Instead of entering −1 a total of 668 times and 1 a total of 58 times, we can enter the following commands in the session window to achieve the same result. (Click on the session window at the top portion of the screen, then click on **Editor**, then click on **Enable Commands**. Enter these commands to create a column consisting of −1 repeated 668 times and 1 repeated 58 times.)

```
SET C3
668(-1), 58(1)
END
```

Now follow the above steps (beginning with step 3) for conducting the sign test, and the results will be as shown below.

```
Sign test of median =  0.00000 versus not = 0.00000


       N  Below  Equal  Above       P  Median
C3   726    668      0     58  0.0000  -1.000
```

The resulting *P*-value of 0.0000 suggests that we reject the null hypothesis of equality of girls and boys. There is sufficient evidence to warrant rejection of the claim that girls and boys are equally likely.

Sign test for testing a claimed value of the median: The textbook also includes an example involving body temperature data. (See Data Set 2 in Appendix B of the textbook for the body temperatures for 12:00 AM on day 2.) We can apply the sign test to test the claim that the *median* is less than 98.6°. Enter the 106 body temperatures in column C3 and follow the steps for doing the sign test, but make these changes in the dialog box entries of Step 6: Enter 98.6 in the box adjacent to "Test median," and select "less than" in the box labeled Alternative. The Minitab results will be as shown below.

```
Sign test of median = 98.60 versus   <   98.60

                N  Below  Equal  Above         P      Median
C3            106     68     15     23    0.0000       98.40
```

The above Minitab display shows that among the 106 body temperatures, 68 are below 98.6, 15 are equal to 98.6, and 23 are above 98.6. The *P*-value of 0.0000 causes us to reject the null hypothesis that the median is at least 98.6. There appears to be sufficient evidence to support the claim that the median body temperature is less than 98.6° F.

13-3 Wilcoxon Signed-Ranks Test

The Triola textbook (excluding *Essentials of Statistics*) describes the Wilcoxon signed-ranks test, and the following definition is given.

Definition

The **Wilcoxon signed-ranks test** is a nonparametric test that uses ranks for these applications:

1. Test a null hypothesis that the population of matched pairs has differences with a median equal to zero.

2. Test a null hypothesis that a single population has a claimed value of the median.

The textbook makes this important point: The Wilcoxon signed-ranks test and the sign test can both be used with sample data consisting of matched pairs, but the sign test uses only the signs of the differences and not their actual magnitudes (how large the numbers are). The Wilcoxon signed-ranks test uses ranks, so the magnitudes of the differences are taken into account. Because the Wilcoxon signed-ranks test incorporates and uses more information than the sign test, it tends to yield conclusions that better reflect the true nature of the data. First we describe the Minitab procedure for conducting a Wilcoxon signed-ranks test, then we illustrate it with an example.

Minitab Procedure for the Wilcoxon Signed-Ranks Test

1. Enter the values of the first variable in column C1.

2. Enter the values of the second variable in column C2.

3. Create column C3 equal to the values of C1 − C2. Accomplish this using either of the following two approaches.

   Session window: Click on the session window at the top portion of the screen, then click on **Editor**, then click on **Enable Commands**. Enter the command:
   LET C3 = C1 − C2.

   Calculator: Click on **Calc**, then **Calculator**. In Minitab's calculator, enter C3 for the "Store result in variable" box, and enter the expression C1 − C2 in the "expression" box. Click **OK** when done.

4. Select **Stat** from the main menu.

5. Select **Nonparametrics** from the subdirectory.

(continued)

6. Select the option of **1-Sample Wilcoxon**.

7. Make these entries in the resulting dialog box:

- **Variables:** For the box labeled Variables, enter C3 (which contains the differences obtained when column C2 is subtracted from column C1).

- **Test Median:** Click on the small circle for "Test median," and, in the adjacent box, enter 0 for a test of the claim that the median is equal to zero.

- **Alterntive:** In the box labeled Alternative, select "not equal" for a test of the claim that the median is equal to zero (so that the alternative hypothesis is that the median is not equal to zero).

- Click **OK**.

Using the same matched data from the table in Section 13-2 of this manual/workbook, the Minitab results for the Wilcoxon signed-ranks test will be as shown below.

Wilcoxon Signed Rank Test: C3

```
Test of median = 0.000000 versus median not = 0.000000

          N for   Wilcoxon            Estimated
     N    Test    Statistic     P      Median
C3   10     9          5.0    0.044    -2.000
```

From these results we see that the test statistic is 5.0 and the *P*-value is 0.044. Because the *P*-value of 0.044 is less than the significance level of 0.05, we reject the null hypothesis that the matched pairs have differences with a median equal to 0. There does appear to be a significant difference between the September weights and the April weights. (Note that the sign test in Section 13-2 led to the conclusion of no difference. By using only positive and negative signs, the sign test did not use the magnitudes of the differences, but the Wilcoxon signed-ranks test was more sensitive to those magnitudes through its use of ranks.)

13-4 Wilcoxon Rank-Sum Test

The Triola textbook (excluding *Essentials of Statistics*) discusses the Wilcoxon rank-sum test and includes the following definition.

Definition

The **Wilcoxon rank-sum test** is a nonparametric test that uses ranks of sample data from two independent populations to test the null hypothesis that the two independent samples come from populations with equal medians. (The alternative hypothesis is the claim that the two populations have different medians, or that the first population has a median greater than the median of the second population, or that the first population has a median less than the median of the second population.)

The Wilcoxon rank-sum test described in the textbook is equivalent to the Mann-Whitney U test, so use Minitab's Mann-Whitney procedure for testing the claim that two independent samples come from populations with populations with equal medians.

Minitab Procedure for the Mann-Whitney Test
(Equivalent to the Wilcoxon Rank-Sum Test)

    1. Enter the two sets of sample data in columns C1 and C2.

    2. Select **Stat** from the main menu.

    3. Select **Nonparametrics**.

    4. Select **Mann-Whitney**.

    5. Make these entries in the dialog box:

- Enter C1 for the first sample and enter C2 for the second sample.

- Enter the confidence level in the indicated box. (A confidence level of 95.0 corresponds to a significance level of $\alpha = 0.05$.)

- Select "alternate: not equal" in the indicated box; that selection refers to the alternative hypothesis, where "not equal" corresponds to a two-tailed hypothesis test.

Shown below are braking distances (in feet) obtained from a random sample of 4-cylinder cars and a random sample of 6-cylinder cars. (Ranks are shown in parentheses.) Also shown are the Minitab results that include a *P*-value of 0.5496. Based on the available sample data, it appears that 4-cylinder cars and 6-cylinder cars have braking distances with the same median.

	Braking Distance (ft)

4-Cylinders	6-Cylinders
136 (12.5)	131 (7.5)
146 (23)	129 (4.5)
139 (15.5)	127 (2)
131 (7.5)	146 (23)
137 (14)	155 (25)
144 (20)	122 (1)
133 (9.5)	143 (18)
144 (20)	133 (9.5)
129 (4.5)	128 (3)
144 (20)	146 (23)
130 (6)	139 (15.5)
140 (17)	136 (12.5)
135 (11)	
$n_1 = 13$	$n_2 = 12$
$R_1 = 180.5$	$R_2 = 144.5$

Mann-Whitney Test and CI: 4 Cyl, 6 Cyl

```
        N   Median
4 Cyl  13   137.00
6 Cyl  12   134.50


Point estimate for ETA1-ETA2 is 2.00
95.3 Percent CI for ETA1-ETA2 is (-6.00,9.00)
W = 180.5
Test of ETA1 = ETA2 vs ETA1 not = ETA2 is significant at 0.5496
The test is significant at 0.5486 (adjusted for ties)
```

13-5 Kruskal-Wallis Test

The Triola textbook (excluding *Essentials of Statistics*) discusses the Kruskal-Wallis test and includes this definition.

Definition

The **Kruskal-Wallis Test** (also called the *H* **test**) is a nonparametric test that uses ranks of simple random samples from three or more independent populations to test the null hypothesis that the populations have the same median. (The alternative hypothesis is the claim that the populations have medians that are not all equal.)

Caution: Using Minitab for a Kruskal-Wallis test is somewhat tricky, because we must enter all of the data in *one column*, then enter the corresponding identifiers in another column. For an example, see the sample data listed below and see the Minitab display showing the data listed in column C1 with identifiers in column C2. The Minitab display shows only the first ten data values, but all of the data must be listed with the same pattern.

If the data are listed in separate columns and you want to stack them as shown below, use **Data/Stack** to combine them into one single column. After selecting **Data/Stack**, you will get a "Stack Columns" dialog box. Enter the columns to be stacked, identify the column where the stacked data will be listed, and identify the column where the corresponding "subscripts" (identifiers) will be listed.

Weights (kg) of Poplar Trees

	None	Fertilizer	Irrigation	Fertilizer and Irrigation
		Treatment		
	0.15	1.34	0.23	2.03
	0.02	0.14	0.04	0.27
	0.16	0.02	0.34	0.92
	0.37	0.08	0.16	1.07
	0.22	0.08	0.05	2.38
n	5	5	5	5
$\bar{x}$	0.184	0.332	0.164	1.334
s	0.127	0.565	0.126	0.859

Stacked Data

↓	C1 Weight	C2-T Treatme
1	0.15	None
2	0.02	None
3	0.16	None
4	0.37	None
5	0.22	None
6	1.34	Fertilizer
7	0.14	Fertilizer
8	0.02	Fertilizer
9	0.08	Fertilizer
10	0.08	Fertilizer

After the sample data have been arranged in a single column with identifiers in another column, we can proceed to use Minitab as follows.

Procedure for the Kruskal-Wallis Test

1. Enter the data values for all categories in one column, and enter the corresponding identification numbers or names in another column.

2. Select **Stat** from the main menu.

3. Select **Nonparametrics**.

4. Select **Kruskal-Wallis**.

5. Make these entries in the dialog box:

   - **Response:** For Response, enter the column with all of the sample data combined.

   - **Factor:** For Factor, enter the column of identification numbers or names.

   - Click **OK**.

Here are the Minitab results obtained by using the Kruskal-Wallis test with the data in the preceding table:

Kruskal-Wallis Test on Weight

```
Treatment      N    Median   Ave Rank        Z
Fert&Irrig     5   1.07000       17.0     2.84
Fertilizer     5   0.08000        7.5    -1.31
Irrigation     5   0.16000        8.5    -0.87
None           5   0.16000        9.0    -0.65
Overall       20                 10.5

H = 8.21   DF = 3   P = 0.042
H = 8.23   DF = 3   P = 0.041   (adjusted for ties)
```

Important elements of the Minitab display include the test statistic of $H = 8.21$ and the P–value of 0.042. We reject the null hypothesis (the samples come from populations with the same median) only if the P–value is small (such as 0.05 or less). Assuming a 0.05 significance level in this case, we reject the null hypothesis of equal medians.

13-6 Rank Correlation

The Triola textbook introduces *rank correlation*, which uses ranks in a procedure for determining whether there is some relationship between two variables.

Definition

The **rank correlation test** (or **Spearman's rank correlation test**) is a nonparametric test that uses ranks of sample data consisting of matched pairs. It is used to test for an association between two variables.

First we describe the Minitab procedure, then we illustrate it with an example.

Minitab Procedure for Rank Correlation

1. Enter the paired data in columns C1 and C2.

2. If the data are already ranks, go directly to step 3. If the data are not already ranks, convert them to ranks by first using Minitab's **Rank** feature. (See Section 13-1 of this manual/workbook.) To convert the data to ranks, make these entries in the Rank dialog box:

   - Enter C1 in the box labeled Rank the data in.

   - Enter C1 in the box labeled Store ranks in.

   - Click **OK**.

   - Enter C2 in the box labeled Rank the data in.

   - Enter C2 in the box labeled Store ranks in.

   - Click **OK**.

3. Select **Stat** from the main menu.

4. Select **Basic Statistics**.

5. Select **Correlation**.

6. Enter C1 C2 in the dialog box, then click **OK**.

Consider the sample data in the table below. The measures of overall quality do not consist of ranks, so we convert them to ranks using Minitab's **Rank** feature, as described in Step 2 above.

Overall Quality Scores and Selectivity Ranks of National Universities

Overall quality	95	63	55	90	74	70	69	86
Selectivity rank	2	6	8	1	4	3	7	5

After replacing the measures of overall quality with their corresponding ranks, we get the table shown below.

Overall Quality Scores and Selectivity Ranks of National Universities

Overall quality	8	2	1	7	5	4	3	6
Selectivity rank	2	6	8	1	4	3	7	5

We now have a table in which both rows are ranks, so we can proceed with Step 3 above. After continuing with Step 3 and the remaining steps, we obtain the following Minitab results.

Correlations: C1, C2

```
Pearson correlation of C1 and C2 = -0.857
P-Value = 0.007
```

The Minitab display shows that the test statistic is -0.857, and the P–value is 0.007. If we use a significance level of 0.05, the P-value of 0.007 is less than the significance level, so we *reject* the null hypothesis of no correlation, and we conclude that there is a correlation.

13-7 Runs Test for Randomness

The Triola textbook (excluding *Essentials of Statistics*) discusses the runs test for randomness and includes these definitions.

Definitions

A **run** is a sequence of data having the same characteristic; the sequence is preceded and followed by data with a different characteristic or by no data at all.

The **runs test** uses the number of runs in a sequence of sample data to test for randomness in the order of the data.

When listing sequences of data to be used for the runs test, the textbook uses both numerical data and qualitative data, but Minitab works with numerical data only, so enter such a sequence using numbers instead of letters. As an illustration, consider the Boston rainfall for Mondays as listed in the Boston rainfall in Data Set 14 of Appendix B from the textbook. If we denote dry days (with 0.00 precipitation) by D's and rainy days (with precipitation amounts greater than 0.00) by R's, we get the following sequence of D's and R's, which we then convert to 0's and 1's as shown below.

DDDDRDRDDRDDRDDDRDDRRRDDDDRDRDRRRDRDDDRDDDRDRDDRDDDR

00001010010010001001110000101011101000100010100 10001

Minitab Procedure for the Runs Test for Randomness

1. In column C1, enter the sequence of *numerical* data. (If the sequence of data consists of two qualitative characteristics, represent the sequence using 0's and 1's.)

2. Select **Stat** from the main menu.

3. Select **Nonparametrics**.

4. Select **Runs** Test.

5. Make these entries in the dialog box:

   • Enter C1 in the box for variables.

   • Click on the button for **Above and below**, and enter a number in the adjacent box. (Choose a number appropriate for the test. If using 0's and 1's, enter 0.5.)

   • Click **OK**.

Suppose that we want to use the runs test to determine whether there is sufficient evidence to support the claim that rain on Mondays is random. Using the preceding sequence of 0's and 1's, we get the following Minitab results. With a *P*-value of 0.140 we fail to reject the null hypothesis of randomness. Rain on Mondays in Boston appears to be random. Big surprise there.

```
Runs test for C1
Runs above and below K = 0.5
The observed number of runs = 30
The expected number of runs = 25.1154
19 observations above K, 33 below
P-value = 0.140
```

CHAPTER 13 EXPERIMENTS: Nonparametric Statistics

In Experiments 13–1 through 13–6, use Minitab's sign test program.

13-1. *Oscar Winners* Listed below are ages of actresses and actors at the times that they won Oscars. The data are paired according to the years that they won. Use a 0.05 significance level to test the claim that there is no difference between the ages of best actresses and the ages of best actors at the time that the awards were presented.

Best Actresses 28 32 27 27 26 24 25 29 41 40 27 42 33 21 35

Best Actors   62 41 52 41 34 40 56 41 39 49 48 56 42 62 29

P–value: _____

Conclusion: _____

_____

13-2. *Airline Fares* Listed below are the costs (in dollars) of flights from New York (JFK) to San Francisco for US Air, Continental, Delta, United, American, Alaska, and Northwest. Use a 0.05 significance level to test the claim that there is no difference in cost between flights scheduled one day in advance and those scheduled 30 days in advance. What appears to be a wise scheduling strategy?

Flight scheduled one day in advance   456   614   628   1088   943   567   536

Flight scheduled 30 days in advance   244   260   264   264   278   318   280

P–value: _____

Conclusion: _____

_____

13-3. *Gender Selection*   The Genetics and IVF Institute conducted a clinical trial of its methods for gender selection. As of this writing, 172 of 211 babies born to parents using the YSORT method were boys. Use a 0.01 significance level to test the claim that the YSORT method is effective in increasing the likelihood of a boy.

P–value: _____

Conclusion: _____

_____

13-4. ***Cheating Gas Pumps*** When testing gas pumps in Michigan for accuracy, fuel-quality enforcement specialists tested pumps and found that 1299 were not accurate (within 3.3 oz when 5 gal is pumped), and 5686 were accurate. Use a 0.01 significance level to test the claim of an industry representative that less than half of Michigan gas pumps are inaccurate.

P–value: _____

Conclusion: _____

_____

13-5. ***Testing for Median Weight of Quarters*** Refer to Data Set 20 in Appendix B for the weights (in g) of randomly selected quarters that were minted after 1964. The quarters are supposed to have a median weight of 5.670 g. Use a 0.01 significance level to test the claim that the median is equal to 5.670 g. Do quarters appear to be minted according to specifications?

P–value: _____

Conclusion: _____

_____

13-6. ***Voltage Levels*** Refer to Data Set 13 in Appendix B for the home voltage levels. The power company (Central Hudson) supplying the power states that the target voltage is 120 V. Use a 0.01 significance level to test the claim that the median voltage is equal to 120 V.

P–value: _____

Conclusion: _____

_____

13–7. ***Sign Test vs. Wilcoxon Signed–Ranks Test*** Repeat Experiment 13-1 by using the Wilcoxon signed-ranks test for matched pairs. Enter the Minitab results below, and compare them to the sign test results obtained in Experiment 13-1. Specifically, how do the results reflect the fact that the Wilcoxon signed-ranks test uses more information?

P–value: _____

Conclusion: _____

_____

Comparison: _____

_____

13–8. ***Sign Test vs. Wilcoxon Signed–Ranks Test*** Repeat Experiment 13-2 by using the Wilcoxon signed-ranks test for matched pairs. Enter the Minitab results below, and compare them to the sign test results obtained in Experiment 13-2. Specifically, how do the results reflect the fact that the Wilcoxon signed-ranks test uses more information?

P–value: _____

Conclusion: _____

_____

Comparison: _____

_____

13-9. ***Coke Contents*** Refer to Data Set 17 in Appendix B for the amounts (in oz) in cans of regular Coke. The cans are labeled to indicate that the contents are 12 oz of Coke. Use a 0.05 significance level with the Wilcoxon signed-ranks test to test the claim that cans of Coke are filled so that the median amount is 12 oz.

P–value: _____

Conclusion: _____

13-10. ***Lengths of Sheet Metal Screws*** Refer to Data Set 19 in Appendix B for the lengths (in inches) of a simple random sample of 50 stainless steel sheet metal screws obtained from those supplied by Crown Bolt, Inc. The screws are packaged with a label indicating 3/4 in. length. Use a 0.05 significance level with the Wilcoxon signed-ranks test to test the claim that the screws have a median equal to 3/4 in. (or 0.75 in.).

P–value: _____

Conclusion: _____

13-11. ***Popes and Monarchs*** Listed below are the numbers of years that U. S. presidents, popes (since 1690), and British Monarchs lived after they were inaugurated, elected, or coronated. As of this writing, the last president is Gerald Ford and the last pope is John Paul II. (The times are based on data from *Computer-Interactive Data Analysis,* by Lunn and McNeil, John Wiley & Sons.) Use a 0.05 significance level with the Wilcoxon rank-sum test to test the claim that the two samples of longevity data from popes and monarchs are from populations with the same median.

Presidents: 10 29 26 28 15 23 17 25 0 20 4 1 24 16 12 4 10 17 16 0 7 24 12 4

18 21 11 2 9 36 12 28 3 16 9 25 23 32

Popes: 2 9 21 3 6 10 18 11 6 25 23 6 2 15 32 25 11 8 17 19 5 15 0 26

Monarchs: 17 6 13 12 13 33 59 10 7 63 9 25 36 15

P–value: _____

Conclusion: _____

_____

13-12. ***Presidents and Popes*** Refer to the longevity data for U.S. presidents and popes in Experiment 13-11. Use a 0.05 significance level with the Wilcoxon rank-sum test to test the claim that the two samples are from populations with the same median.

P–value: _____

Conclusion: _____

_____

13-13. ***Appendix B Data Set: Cigarettes***   Refer to Data Set 4 in Appendix B for the amounts of nicotine (in mg per cigarette) in the sample of king size cigarettes, which are non-filtered, non-menthol, and non-light, and for the amounts of nicotine in the 100 mm cigarettes, which are filtered, non-menthol, and non-light. Use a 0.01 significance level with the Wilcoxon rank-sum test to test the claim that the median amount of nicotine in the non-filtered king size cigarettes is greater than the median amount of nicotine in the 100 mm filtered cigarettes.

P–value: _____

Conclusion: _____

_____

13-14. ***Appendix B Data Set: Cigarettes***   Refer to Data Set 4 in Appendix B for the amounts of tar in the sample of king size cigarettes, which are non-filtered, non-menthol, and non-light, and for to the amounts of tar in the 100 mm cigarettes, which are filtered, non-menthol, and non-light. Use a 0.01 significance level with the Wilcoxon rank-sum test to test the claim that the median amount of tar in the non-filtered king size cigarettes is greater than the median amount of tar in the 100 mm filtered cigarettes.

P–value: _____

Conclusion: _____

_____

13–15. ***Do All Colors of M&Ms Weigh the Same?*** Refer to the M&M data set in Appendix B from the textbook. Use a 0.05 significance level with the Kruskal-Wallis test to test the claim that the weights of M&Ms have the same median for each of the six different color populations.

P–value: _____

Conclusion: _____

_____

13-16. ***Appendix B Data Set: Nicotine in Cigarettes*** Refer to Data Set 4 in Appendix B and use the amounts of nicotine (mg per cigarette) in the king size cigarettes, the 100 mm menthol cigarettes, and the 100 mm non-menthol cigarettes. The king size cigarettes are non-filtered, non-menthol, and non-light. The 100 mm menthol cigarettes are filtered and non-light. The 100 mm non-menthol cigarettes are filtered and non-light. Use a 0.05 significance level with the Kruskal-Wallis test to test the claim that the three categories of cigarettes yield the same median amount of nicotine.

P–value: _____

Conclusion: _____

_____

13-17. ***Appendix B Data Set: Tar in Cigarettes*** Refer to Data Set 4 in Appendix B and use the amounts of tar (mg per cigarette) in the three categories of cigarettes described in Exercise 11. Use a 0.05 significance level with the Kruskal-Wallis test to test the claim that the three categories of cigarettes yield the same median amount of tar.

P–value: _____

Conclusion: _____

_____

13-18. ***Judges of Marching Bands*** Two judges ranked seven bands in the Texas state finals competition of marching bands (Coppell, Keller, Grapevine, Dickinson, Poteet, Fossil Ridge, Heritage), and their rankings are listed below (based on data from the University Interscholastic League). Test for a correlation between the two judges. Do the judges appear to rank about the same or are they very different?

Band	Cpl	Klr	Grp	Dck	Ptt	FR	Her
First Judge	1	3	4	7	5	6	2
Second Judge	6	4	5	1	3	2	7

P–value: _____

Conclusion: _____

_____

13-19. ***Judges of Marching Bands*** In the same competition described in Experiment 13-18, a third judge ranked the bands with the results shown below. Test for a correlation between the first and third judges. Do the judges appear to rank about the same or are they very different?

Band	Cpl	Klr	Grp	Dck	Ptt	FR	Her
First Judge	1	3	4	7	5	6	2
Third Judge	3	4	1	5	7	6	2

P–value: _____

Conclusion: _____

_____

13-20. ***Plasma TVs*** *Consumer Reports* magazine tested large plasma TVs. The table below shows the rankings of TVs by overall quality score and cost. High values are given low ranks, so the TV with a quality rank of 1 is the TV with the highest quality score, and a TV given a cost rank of 1 is the most expensive TV. Test for a correlation. Based on these results, can you expect to get higher quality by purchasing a more expensive plasma TV?

Quality	1	2	3	4	5	6	7	8	9	10
Cost	2	3	6	1	10	4	9	5	8	7

P–value: _____

Conclusion: _____

_____

13-21. ***LCD TVs*** *Consumer Reports* magazine tested LCD televisions. The table below shows the overall quality score and cost in hundreds of dollars. Test for a correlation. Based on these results, can you expect to get higher quality by purchasing a more expensive LCD television?

Quality	74	71	68	65	63	62	59	57	57	53	51
Cost	27	30	38	23	20	13	27	23	14	13	20

P–value: _____

Conclusion: _____

_____

13-22. ***Oscar Winners*** Listed below are the genders of the younger winner in the categories of Best Actor and Best Actress. for recent and consecutive years. Do the genders of the younger winners appear to occur randomly?

F F F M M F F F F F F M M M M M

P–value: _____

Conclusion: _____

13-23. ***Testing for Randomness of Presidential Election Winners*** The political parties of the winning candidates for a recent sequence of presidential elections are listed below, where D denotes Democratic party and R denotes Republican party. Does it appear that we elect Democrat and Republican candidates in a random sequence?

R R D R D R R R R D D R R R D D D

D D R R D D R R D R R R D D R R

P–value: _____

Conclusion: _____

13-24. ***Baseball World Series Victories*** Test the claim that the sequence of World Series wins by American League and National League teams is random. Given below are recent results, with American League and National League teams represented by A and N, respectively.

A N A N N N A A A A N A A A A N A N N A A N N A A A A A N A N

N A A A A A N A N A N A N A A A A A A A N N A N A N N A A N

N N A N A N A N A A A N N A A N N N N A A A N A N A N A A A

N A N A A A N A N A A N A

P–value: _____

Conclusion: _____

14

Statistical Process Control

14-1 Run Charts

In the Statistical Process Control chapter (Chapter 14) in the Triola textbook (excluding *Essentials of Statistics*) we define **process data** to be data arranged according to some time sequence, such as the data in the table below. At the Altigauge Manufacturing Company, four altimeters are randomly selected from production on each of 20 consecutive business days, and the table lists the *errors* (in feet) when they are tested in a pressure chamber that simulates an altitude of 1000 ft. The error measurements are displayed in the table with bold text. On day 1, for example, the actual altitude readings for the four selected altimeters are 1002 ft, 992 ft, 1005 ft, and 1011 ft, so the corresponding errors (in feet) are 2, −8, 5, and 11. Those four errors have a mean of 2.50 ft, a median of 3.5 ft, a range of 19 ft, and a standard deviation of 7.94 ft.

Aircraft Altimeter Errors (in feet)

Day						Mean	Median	Range	St. Dev.
1	2	-8	5	11		2.50	3.5	19	7.94
2	-5	2	6	8		2.75	4.0	13	5.74
3	6	7	-1	-8		1.00	2.5	15	6.98
4	-5	5	-5	6		0.25	0.0	11	6.08
5	9	3	-2	-2		2.00	0.5	11	5.23
6	16	-10	-1	-8		-0.75	-4.5	26	11.81
7	13	-8	-7	2		0.00	-2.5	21	9.76
8	-5	-4	2	8		0.25	-1.0	13	6.02
9	7	13	-2	-13		1.25	2.5	26	11.32
10	15	7	19	1		10.50	11.0	18	8.06
11	12	12	10	9		10.75	11.0	3	1.50
12	11	9	11	20		12.75	11.0	11	4.92
13	18	15	23	28		21.00	20.5	13	5.72
14	6	32	4	10		13.00	8.0	28	12.91
15	16	-13	-9	19		3.25	3.5	32	16.58
16	8	17	0	13		9.50	10.5	17	7.33
17	13	3	6	13		8.75	9.5	10	5.06
18	38	-5	-5	5		8.25	0.0	43	20.39
19	18	12	25	-6		12.25	15.0	31	13.28
20	-27	23	7	36		9.75	15.0	63	27.22

A *run chart* is a sequential plot of *individual* data values over time. A run chart can be generated as follows.

Minitab Procedure for Generating a Run Chart

1. Enter all of the data (in sequence) in column C1.

2. Select the main menu item of **Stat**.

3. Select the subdirectory item of **Quality Tools**.

4. Select the option of **Run Chart**.

5. Make these entries in the dialog box:

   • Select **Single column** and enter C1 in the adjacent box.

   • Enter 1 for the subgroup size (because we want *individual* values plotted).

   • Click **OK**.

Using the process data in the preceding table, the above procedure will result in this run chart:

Examine the above run chart and note that it reveals this problem: As time progresses from left to right, the heights of the points appear to show a pattern of increasing variation. See how the points at the left fluctuate considerably less than the points farther to the right. The Federal Aviation Administration regulations require errors less than 20 ft (or between 20 ft and −20 ft), so the altimeters represented by the points at the left are OK, whereas several of the points farther to the right correspond to altimeters not meeting the required specifications. It appears that the manufacturing process started out well, but deteriorated as time passed. If left alone, this manufacturing process will cause the company to go out of business. That is, the run chart suggests that the process is not **statistically stable** (or **within statistical control**) because it has a pattern of increasing variation.

14-2 *R* Charts

The textbook describes *R* charts as sequential plots of ranges. Using the data from the preceding table, for example, the *R* chart is a plot of the ranges 19, 13, 15, . . . , 63. *R* charts are used to monitor the *variation* of a process. The procedure for obtaining an *R* chart is as follows.

Minitab Procedure for Creating *R* Charts

1. Enter all of the data in sequential order in column C1.

2. Select **Stat** from the main menu.

3. Select **Variable Charts for** Subgroups.

4. Select the option of **R**.

5. Make these entries in the dialog box:

   • Enter C1 in the box for the column containing all observations.

   • Enter a subgroup size of 4 in the indicated box. (We use 4 here because the table of altimeter errors include 4 observations each day.)

   • Click the **R options** button, click on the Estimate tab, and click on the **Rbar** button. Click on **Estimate**, then click on the small circle labeled **Rbar**. Click **OK**.)

Using the data in the preceding table, the Minitab display will be as shown below.

We can interpret the above R chart by applying these three out-of-control criteria given in the textbook:

1. There is no pattern, trend, or cycle that is obviously not random.

2. No point lies beyond the upper or lower control limits.

3. There are not 8 consecutive points all above or all below the center line.

Analyzing the above R chart leads to the conclusion that the process *is out of statistical control* because the second criterion is violated: There is a point beyond the upper control limit. Also, there is a pattern of an upward trend.

14-3 $\overline{x}$ **Charts**

We can now proceed to use the same altimeter error data listed in the table from Section 14-1 of this manual/workbook to create an $\overline{x}$ chart. The textbook explains that an $\overline{x}$ chart is used to monitor the *mean* of the process. It is obtained by plotting the sample means. Follow these steps to generate an $\overline{x}$ chart.

Minitab Procedure for Generating an $\overline{x}$ Chart

1. Enter all of the data in sequential order in column C1.

2. Select **Stat** from the main menu.

3. Select **Variable Charts for** Subgroups.

4. Select the option of **Xbar**.

5. Make these entries in the dialog box:

- Enter C1 in the box for the column containing all observations.

- Enter a subgroup size of 4 in the indicated box. (We use 4 here because the data in the table of altimeter errors includes 4 observations each day.)

- Click **Xbar options**, click the **Estimate** tab, and click on the **Rbar** button.

Using the data in table of altimeter errors, the Minitab display will be as shown below.

We can interpret the $\bar{x}$ chart by applying the three out-of-control criteria given in the textbook. We conclude that the mean in this process is *out of statistical control* because there does appear to be a pattern of an upward trend, and the last of the following conditions is also violated:

1. There is no pattern, trend, or cycle that is obviously not random.

2. No point lies beyond the upper or lower control limits.

3. There are not 8 consecutive points all above or all below the centerline.

14-4 *p* Charts

A control chart for attributes (or *p* chart) can also be constructed by using the same procedure for *R* charts and $\bar{x}$ charts. A *p* chart is very useful in monitoring some process proportion, such as the proportions of defects over time. As an example, the Altigauge Manufacturing Company produces altimeters in batches of 100, and each altimeter is tested and determined to be acceptable or defective. Listed below are the numbers of defective altimeters in successive batches of 100.

Defects: 2 0 1 3 1 2 2 4 3 5 3 7

Follow these steps to generate a Minitab *p* chart:

Minitab Procedure for Generating a *p* Chart

1. Enter the numbers of defects (or items with any particular attribute) in column C1.

2. Select the main menu item of **Stat**.

3. Select the subdirectory item of **Control Charts**.

4. Select the option of **Attribute** Charts.

5. Make these entries in the dialog box:

   - Enter C1 in the box labeled Variable.

   - Enter the subgroup size. (For example, if C1 consists of the numbers of defects per batch and each batch consists of 100 altimeters, then enter 100 for the subgroup size.)

   - Click **OK**.

The Minitab display will be as shown below. We can interpret the control chart for *p* by considering the three out-of-control criteria listed in the textbook. Using those criteria, we conclude that this process is out of statistical control for this reason: There appears to be an upward trend. The company should take immediate action to correct the increasing proportion of defects.

CHAPTER 14 EXPERIMENTS: Statistical Process Control

Constructing Control Charts for Aluminum Cans *Experiments 1 and 2 are based on the axial loads(in pounds) of aluminum cans that are 0.0109 in. thick, as listed in the "Axial Loads of Aluminum Cans" data set in Appendix B of the textbook. An axial load of a can is the maximum weight supported by its side, and it is important to have an axial load high enough so that the can isn't crushed when the top lid is pressed into place. The data are from a real manufacturing process, and they were provided by a student who used an earlier version of this book.*

14–1. **R Chart**   On each day of production, seven aluminum cans with thickness 0.0109 in. were randomly selected and the axial loads were measured. The ranges for the different days are listed below, but they can also be found from the values given in Appendix B of the textbook. (The Minitab worksheet is *CANS*.) Construct an *R* chart and determine whether the process variation is within statistical control. If it is not, identify which of the three out-of-control criteria lead to rejection of statistically stable variation.

78 77 31 50 33 38 84 21 38 77 26 78 78
17 83 66 72 79 61 74 64 51 26 41 31

14–2. $\bar{x}$ **Chart**   On each day of production, seven aluminum cans with thickness 0.0109 in. were randomly selected and the axial loads were measured. The means for the different days are listed below, but they can also be found from the values given in Appendix B of the textbook. (The Minitab worksheet is *CANS*.) Construct an $\bar{x}$ chart and determine whether the process mean is within statistical control. If it is not, identify which of the three out-of-control criteria lead to rejection of statistically stable variation.

252.7 247.9 270.3 267.0 281.6 269.9 257.7 272.9 273.7 259.1
275.6 262.4 256.0 277.6 264.3 260.1 254.7 278.1 259.7 269.4
266.6 270.9 281.0 271.4 277.3

Energy Consumption. In Experiments 14-3 through 14-5, refer to Data Set 12 in Appendix B and use the amounts of electricity consumed (in kWh) in the author's home. Let each subgroup consist of the six amounts within the same year, so that there are eight subgroups with six amounts in each subgroup.

14-3. **Energy Consumption: R Chart**   Let each subgroup consist of the 6 values within a year. Construct an *R* chart and determine whether the process variation is within statistical control. If it is not, identify which of the three out-of-control criteria lead to rejection of statistically stable variation.

14-4.   ***Energy Consumption:*** $\bar{x}$ ***Chart*** Let each subgroup consist of the 6 values within a year. Construct an $\bar{x}$ chart and determine whether the process mean is within statistical control. If it is not, identify which of the three out-of-control criteria lead to rejection of a statistically stable mean.

14-5.   ***Energy Consumption: Run Chart*** Construct a run chart for the 48 values. Does there appear to be a pattern suggesting that the process is not within statistical control?

Energy Consumption. *In Exercises 14-6 through 14-8, refer to Data Set 13 in Appendix B and use the measured voltage amounts for the power supplied directly to the author's home. Let each subgroup consist of the five amounts within the business days of a week, so the first five voltages constitute the first subgroup, the second five voltages constitute the second subgroup, and so on. The result is eight subgroups with five values each.*

14-6.   ***Home Voltage:*** $\bar{x}$ ***Chart*** Using subgroups of five voltage amounts, construct an $\bar{x}$ chart and determine whether the process mean is within statistical control. If it is not, identify which of the three out-of-control criteria lead to rejection of a statistically stable mean.

14-7.   ***Home Voltage: Run Chart*** Construct a run chart for the 40 voltage amounts. Does there appear to be a pattern suggesting that the process is not within statistical control?

14-8.   ***Home Voltage: R Chart*** Using subgroups of five voltage amounts, construct an R chart and determine whether the process variation is within statistical control. If it is not, identify which of the three out-of-control criteria lead to rejection of statistically stable variation.

14-9.   ***p Chart for Defective Defibrillators*** Consider a process that includes careful testing of each manufactured defibrillator. Listed below are the numbers of defective defibrillators in successive batches of 10,000. Construct a control chart for the proportion p of defective defibrillators and determine whether the process is within statistical control. If not, identify which of the three out-of-control criteria apply.

Defects: 20 14 22 27 12 12 18 23 25 19 24 28 21 25 17 19 17 22 15 20

14-10. ***p Chart for Defective Defibrillators*** Repeat Experiment 14-9 assuming that the size of each batch is 100 instead of 10,000. Compare the control chart to the one found for Experiment 14-9. Comment on the general quality of the manufacturing process described in Experiment 14-9 compared to the manufacturing process described in this exercise.

14-11. ***p Chart for College Enrollment*** In each of 15 recent and consecutive years, 1000 high school completers were randomly selected and the number who enrolled in college was determined, with the results listed below. Construct a control chart for the proportion p of high school completers who enroll in college. Does the p chart indicate that such college enrollments are high enough? (The values are based on data from the U. S. National Center for Education Statistics, and they are the most recent values available at the time of this writing.)

601 625 619 626 619 619 650 670 656 629 633 618 652 639 667

14-12. ***p Chart for Violent Crimes*** In each of 15 recent and consecutive, 100,000 people in the U. S. were randomly selected and the number who were victims of violent crime was determined, with the results listed below. Does the rate of violent crime appear to exhibit acceptable behavior? (The values are based on data from the U. S. Department of Justice, and they are the most recent values available at the time of this writing.)

730 758 758 747 714 685 637 611 568 523 507 505 474 476 486

Minitab Release 15
Important Menu Items

File Edit Data Calc Stat Graph Editor Tools Window Help

Copy Cells
Paste Cells

Enable Commands

New	**Basic Statistics**[2]	**Scatterplot**
Open Worksheet	**Regression**	**Histogram**
Save Current Worksheet As	**ANOVA**	**Dotplot**
Other Files[1]	**Control Charts**	**Stem-and-Leaf**
Print Worksheet	**Quality Tools**	**Boxplot**
Print Session Window	**Nonparametrics**	**Bar Chart**
Exit	**EDA**	**Pie Chart**
		Time Series Plot

Stack	**Calculator**
Sort	**Column Statistics**
Rank	**Row Statistics**
	Random Data
	Probability Distributions

[1] **Other Files:**   Use for importing and exporting text files.

[2] **Basic Statistics:**   **Descriptive Statistics**
Confidence intervals and hypothesis tests for one or two proportions
Confidence intervals and hypothesis tests for one or two means
Confidence intervals and hypothesis tests for matched pairs
Confidence intervals and hypothesis tests for one or two variances
Linear correlation
Test for normality

Retrieving Data from the CD-ROM

The CD-ROM included with the Triola textbook has worksheets for the data in Appendix B of the textbook. Appendix B in the textbook lists the worksheet names, and they are also listed on the following page. Here is the procedure for retrieving a worksheet from the CD-ROM:

1.   Click on Minitab's main menu item of **File**, then click on **Open Worksheet**.

2.   You will get a dialog box like the one shown below. In the "Look in" box at the top, select the location of the stored worksheets. For example, if the CD-ROM is in drive D and you want to open the Minitab worksheet WORDS, do this:

   - In the "Look in" box, select drive **D** (or whatever drive contains the CD-ROM).
   - Double click on the folder **App B Data Sets**.
   - Double click on the folder **Minitab**.
   - Click on **WORDS** (or any other worksheet that you want), as shown below.

3.   Click on the **Open** bar located in the lower right portion of the dialog box.

4.   You will get a message that a copy of the contents of the file will be added to the current project. Click **OK**. The columns of data will now be in the Minitab worksheet.

Appendix B: Data Sets — Minitab Worksheet Name

Data Set	Description	Minitab Worksheet Name
Data Set 1:	Health Exam Results	MHEALTH FHEALTH
Data Set 2:	Body Temperatures of Healthy Adults	BODYTEMP
Data Set 3:	Freshman 15 Data	FRESH15
Data Set 4:	Cigarette Tar, Nicotine, and Carbon Monoxide	CIGARET
Data Set 5:	Passive and Active Smoke Measurements	COTININE
Data Set 6:	Bear Measurements	BEARS
Data Set 7:	Alcohol and Tobacco Use in Animated Children's Movies	CHMOVIE
Data Set 8:	Word Counts by Males and Females	WORDS
Data Set 9:	Movie Data	MOVIES
Data Set 10:	NASA Space Transport System Data	NASA
Data Set 11:	Forecast and Actual Temperatures	WEATHER
Data Set 12:	Electricity Consumption of a Home	ELECTRIC
Data Set 13:	Voltage Measurements from a Home	VOLTAGE
Data Set 14:	Rainfall in Boston for One Year	BOSTRAIN
Data Set 15:	Old Faithful Geyser Measurements	OLDFAITH
Data Set 16:	Car Measurements	CARS
Data Set 17:	Weights and Volumes of Cola	COLA
Data Set 18:	M&M Plain Candy Weights	M&M
Data Set 19:	Screw Lengths	SCREWS
Data Set 20:	Coin Weights	COINS
Data Set 21:	Axial Loads of Aluminum Cans	CANS
Data Set 22:	Weights of Discarded Garbage for One Week	GARBAGE
Data Set 23:	Home Sales	HOMES
Data Set 24:	FICO Credit Rating Scores	FICO

Index